Praise for
<LOW-CODE/NO-C

"I read all of Phil's books. As a software developer coming of age in the 1980s and early 1990s, I fondly remember tools like DataFlex, Microsoft Access, and Clarion. Phil shows how today's citizen developers can use low-code/no-code software in amazing ways while honoring the trajectory that got us here."

> **—BRAD FELD**, partner at Foundry, cofounder of Techstars,
> author, runner, reader, hermit, and nerd

"*Low-Code/No-Code* could not have arrived at a better time. Simon systematically identifies the most pressing issues that tech execs, app developers, and IT professionals face—and adeptly shows how to resolve them."

> **—GERALD C. KANE,** C. Herman and Mary Virginia Terry Chair
> in Business Administration, University of Georgia and
> coauthor of *The Transformation Myth*

"Take it from me: The demand for new internal apps and tech has only intensified, but software engineers are more difficult to recruit and retain than ever. Something has to give. Fortunately, *Low-Code/No-Code* shows the path forward. New tools let subject matter experts become citizen developers and shoulder a good deal of the development burden. Get this book."

> **—PAUL MAYA,** CTO, United States Tennis Association

"If you haven't heard of low-code/no-code yet, rest assured: You will soon. The revolution is here, but we need a guide to effectively navigate it. Phil Simon's latest book on workplace technology is just the ticket. Ignore this well-researched, exceptionally relevant, and downright inspiring book at your own peril."

> **—MATT WADE**, consultant and author

"A delightfully easy read that left me informed and inspired. An essential text for any curious self-starter."

"Once again, Phil Simon makes a critical and emerging technology trend accessible. Chief executives, department heads, and project managers are just some of the folks who will benefit from this opportune and important book."

"In *Low-Code/No-Code,* Phil Simon shares how the emergence of the new way of coding came to be. He emphatically yet elegantly states why it's imperative for both large and small organizations to start riding this wave today."

"Monkeys randomly hitting keys on a typewriter for an infinite amount of time will almost surely produce Hamlet. Now imagine what smart people equipped with amazing tools will produce. For a more concrete way to understand the power of low-code/no-code, read this important book."

"As of now, the size of the low-code/no-code market is $15 billion—hefty, but a pittance compared to what it will soon become. I firmly believe that the next trillion-dollar company will be one of the existing no-code vendors that Phil Simon lists in his compelling new book. It is a must-read if you are interested in the future of software development."

<LOW-CODE/
NO-CODE>

<LOW-CODE/ NO-CODE>

Citizen Developers and the Surprising Future of Business Applications

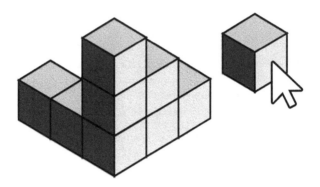

>_PHIL SIMON

Award-winning author of *Reimagining Collaboration*
and *Project Management in the Hybrid Workplace*

RACKET

Arizona

For information about this title or to order other books and/or electronic
media, contact the publisher:

Racket Publishing | www.racketpublishing.com

ISBNs
9798985814736 (paperback)
9798985814750 (hardcover)
9798985814743 (ebook)

Printed in the United States of America

Cover design: Luke Fletcher | www.fletcherdesigns.com
Interior design: Jessica@dezinermama.com

Project Management in the Hybrid Workplace

Reimagining Collaboration: Slack, Microsoft Teams, Zoom, and the Post-COVID World of Work

Zoom For Dummies

Agile: The Insights You Need from Harvard Business Review (contributor)

Slack For Dummies

Analytics: The Agile Way

Message Not Received: Why Business Communication Is Broken and How to Fix It

The Visual Organization: Data Visualization, Big Data, and the Quest for Better Decisions

Too Big to Ignore: The Business Case for Big Data

The Age of the Platform: How Amazon, Apple, Facebook, and Google Have Redefined Business

The New Small: How a New Breed of Small Businesses Is Harnessing the Power of Emerging Technologies

The Next Wave of Technologies: Opportunities in Chaos

Why New Systems Fail: An Insider's Guide to Successful IT Projects

To the eternal spirit of Taylor Hawkins

"The future of coding is no coding at all."

—CHRIS WANSTRATH, CO-FOUNDER AND EX-CEO, GITHUB

>_Contents

>_List of Figures and Tables

Figure 1.1: The Experience Chasm in Workplace Tech

Figure 1.2: Technology Tradeoffs

Figure 1.3: The War for Tech and Data Talent Is Real

Figure 2.1: Traditional Applications and Systems

Table 2.1: Popular Categories of Business Applications

Table 2.2: Additional Categories of Business Systems

Table 2.3: Comparing the Mainstream Software Deployment Approaches

Table 3.1: The Generations of Programming Languages

Figure 3.1: Microsoft FrontPage User Interface

Figure 3.2: Primary Tools by Type of Coder

Figure 3.3: LC/NC Tools: Then and Now

Figure 3.4: Building Modern Business Applications

Figure 3.5: Divi WordPress Theme

Figure 4.1: Revenue From Low-Code Development Technologies (Billions of US Dollars)

Table 5.1: Major Differences Between Traditional Software Developers and Citizen Developers

Figure 5.1: 2023 Projected Developer Breakdown

Figure 5.2: The Virtuous Cycle of Citizen Developers

Figure 5.3: Investment Skyrockets in LC/NC While BPMS Drops Significantly

<Part I>

Application Development in the Here and Now

Chapter 1

Workplace Tech:
The Struggle Is Real

"Where you stand depends on where you sit."

—RUFUS MILES

For decades, our relationship with workplace systems, applications, and technology has been complicated.

In October 2018, the high-end professional services firm PricewaterhouseCoopers surveyed more than 12,000 employees across different countries and industries.[1] As a lot, rank-and-file workers chafed at the tools their organizations required them to use. For example, nearly three in four respondents reported knowing of technologies that would improve the quality of their output.

A little more than half felt that their leadership chose workplace tech with their employees in mind. Perhaps not surprisingly, the folks in the corner offices sang a far more sanguine tune. Figure 1.1 displays the chasm between the two cohorts.

The Experience Chasm in Workplace Tech

Percentage of respondents who agree with the statement: My company pays attention to people's needs when introducing new technologies.

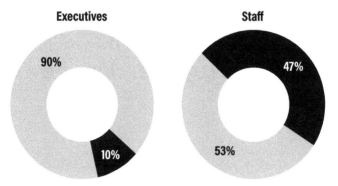

Figure 1.1: The Experience Chasm in Workplace Tech
Source: PricewaterhouseCoopers

In a word, *wow*.

At least there's an upside. (There's nowhere to go but up, right?) In its report, PwC correctly notes that the status quo affords plenty of opportunity for senior leaders interested in improving their employees' well-being. The top brass would do well to finally address the critical and longstanding deficit in workplace tech. Sure, the required investments in new applications and technologies would be sizeable, but so would the potential returns. Employers could increase employee job satisfaction, performance, and even retention.

Would the juice be worth the squeeze?

In theory, the answer was obvious.

Isolating Cause and Effect

In practice, however, things were murkier. After all, the real world is a messy place with myriad factors at play. It's nearly impossible to ascertain if an independent variable (increased tech spending) affects a dependent one (employee performance), never mind to what extent. The arrow can go both ways.

In an alternate universe, researchers could use the controlled environment of a laboratory to answer these interdependent questions. Experiments, placebos, random assignments, double-blind studies, and other techniques would definitively isolate cause and effect and, in the process, provide conclusive, inarguable answers.

Fortunately, all hope is not lost. Academics, researchers, and social scientists have for decades embraced natural experiments to find solutions to thorny real-world problems. They routinely study and analyze real-life situations to ascertain the relationships among variables. Their job isn't easy.

One such natural experiment would arrive not long after the publication of the PwC survey in the form of COVID-19. In this limited and perverse sense, the pandemic could shed light on the effects of massive corporate expenditures in workplace tech. Specifically, how would they affect employee performance, satisfaction, and retention?

We were about to find out.

Gartner predicted that global business IT spending in the US alone in 2021 would hit $3.8 trillion, an increase of 4 percent from 2020.[2] No doubt that part of the uptick stemmed from the pandemic. In November 2020, KPMG reported that businesses had spent an estimated $15 billion extra *per week* on technology to enable remote work.[3]

Those numbers certainly qualify as significant cabbage, but was the money well spent? What was the return on investment (ROI)?

The word *disappointing* comes to mind.

A June 2022 study found the following:

> Ninety-one percent of employees reported being frustrated with the applications they use in their current positions.
> Seventy-one percent of leaders acknowledge that employees will consider looking for a new job if their current employer doesn't provide access to the tools, technology, or information they need to do their jobs well.[4]

According to a separate 2021 survey, one in five employees said their existing workplace technology made their job harder.[5]

Who's to Blame for the Status Quo?

Throw shade at your IT department if you want, but it—and its employees—are often convenient scapegoats. Rapidly rolling out new applications and systems isn't easy under normal circumstances, to say nothing about pandemics. Regardless of what management gurus claim, digital transformation—a piece of business jargon that I despise—is a tough row to hoe.

In reality, resource-related challenges are old hat. They've been plaguing IT departments since long before the pandemic. Exhibit A: In April 2018, Salesforce Research released its Enterprise Technology Trends report.[6] Among its most interesting findings: 72 percent of IT leaders claimed they lacked the time and resources to work on their own strategic projects. Ever-increasing project backlogs prevented them from focusing on their critical priorities. Again, the company gathered these statistics *before* COVID-19, and the Great Resignation shook many IT departments—and the organizations behind them—to their foundations.

More than a year into the pandemic, a decent chunk of employers struggled to stay afloat, let alone be profitable. In June 2021, Accenture reported that 28 percent of firms were operating "without proper tools and processes at scale." What's more, one in five organizations claimed that their backlog exceeded fifty initiatives.[7]

Ouch.

Now, all departments within any given enterprise matter in the abstract. It turns out, however, that some groups are *more* important—a point that COVID-19 drove home.

The Pandemic Validated the Importance of Workplace Tech

Starting in March 2020, workplace technology went from *important* to *downright essential.* The reason is simple: Under the extraordinary circumstances of COVID-19, millions of employees simply couldn't work without lots of new tech.

Slack cofounder and CEO Stewart Butterfield was an early media darling on how to navigate our new normal. One of his frequent talking points is particularly salient here: Although employees couldn't meet in person during the days of lockdown, they were able to remain productive because of the communication and collaboration tools at their disposal. (And, yes, Slack is one of them, although now it's part of Salesforce.)

The data supports Butterfield's assertion. HR and workplace benefits consulting firm Mercer surveyed 800 employers in the months following the outbreak of COVID-19. Ninety-four percent of respondents said that employee productivity was the same or higher than before the pandemic *even when they worked remotely.**

Flip the scenario, though. Imagine if we'd worked in our usual offices and met in person, but we couldn't communicate via sophisticated digital technologies. The pandemic's work-related outcomes—equivalent worker productivity and relatively minimal disruption—almost certainly would have been different.

---

* The survey lives behind a paywall, but the company made an infographic available for free at https://tinyurl.com/mercerRPM.

When it came to work, technology saved us.

There. I said it.

Survey results confirm that information technology and the people who lead it are now more likely to be sitting at the big-boy table. Consider the 2020 Harvey Nash/KPMG CIO Survey of 4,219 global IT leaders from a variety of industries.[8] More than three in five reported that the pandemic had *increased* their influence within their organizations and with their colleagues. More than four in five expected their budgets and headcounts to grow in the next year. Upon hearing the news, I suspect most of them reacted with the words, "About freakin' time."

As I've been saying for more than a decade, *all* companies are tech companies. Some just haven't realized it yet.[*] (Cue Marc Andreessen's quote about software eating the world.) The pandemic shed light on many things; the importance of technology for any contemporary business is just one of them.

Understanding the Dual Nature of Contemporary IT

One could write a lengthy book about the many competing and even conflicting demands that IT leaders and departments routinely face. Many people have. At a high level, their challenges stem from one simple reality: The notion of a monolithic IT department is an antiquated one. Contemporary IT represents two related but distinct groups. And this brings us to the world of DevOps.

Its origins trace back thirty years, but DevOps has only taken off in the past eight. (Google Trends makes me seem smarter than I am.[†])

---

[*] That sentence has adorned the home page of my website for a decade.
[†] See for yourself at https://tinyurl.com/dev-phil-ops.

This portmanteau fuses two essential technical disciplines: development and IT operations. In theory, the organization that has embraced DevOps is facing reality. Its management has recognized that employees who build new applications and launch new features typically don't share the same priorities as their counterparts tasked with upgrading and maintaining existing systems and applications. Developers get jazzed about shiny new things; IT security analysts worry about leaks, hacks, malware, and ransomware—as they damn well should.

In October 2021, Rackspace Technology surveyed 1,420 IT professionals.[9] The results may surprise you, but grizzled industry types just nodded their heads:

> A full half of global IT leaders reported that they weren't "fully confident" in their ability to respond to an increasing array of intricate threats.
> Perhaps most alarmingly, 86 percent of respondents revealed that their organizations lack the necessary skills, expertise, and resources.

At least the IT bigwigs aren't alone. Consider the results of a February 2020 McKinsey Global Survey. Nearly nine in ten executives reported experiencing skill gaps in the workforce or anticipated them within a few years.[10]

Yikes.

Economics is the study of scarcity. Paul Samuelson called it "a choice between alternatives all the time." One doesn't need to be John Maynard Keynes or Adam Smith, however, to appreciate the fundamental tradeoff between developers and operations folks. By definition, a dollar for one group means one fewer dollar for the other. (A notable exception: JPMorgan Chase CEO Jamie Dimon has said that his company's cybersecurity is effectively unlimited. He previously authorized more than $600 million in *a single year* to protect his customers' assets and information.[11])

As Figure 1.2 shows, the resource tradeoff is real.

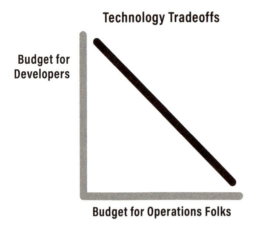

Figure 1.2: Technology Tradeoffs

Bottom line: Organizations spend more on operations, maintenance, and security. As a result, there's less money to allocate to proper developers.

Before moving on, let's take a time-out.

You may have read the first few pages of this chapter and yawned. (I can take it.) Why is this information relevant? More concretely, say that you work at a one-hundred-person entertainment company with a single full-time IT employee. Does all of this DevOps stuff *really* affect you?

Trust me. It does.

Regardless of whom you call when your computer doesn't work, the odds are that the dual nature of IT function profoundly influences how you and your colleagues do your jobs. Consider the following scenarios:

> You found a bug in one of your company's homegrown systems. Six months later, it's still there mocking you.
> A ransomware hack prevented you from accessing critical systems and information for a month.
> After weeks of back-and-forth to set up a meeting to discuss new enhancements to key systems, the head of IT cancels because she needs to fight another fire.
> A single IT employee unexpectedly leaves. Who will mind the store as management frantically searches for a suitable replacement?

I suspect that you're getting my drift. If these circumstances eerily describe your current sitch, at least you can solace in the fact that you're not alone. Over the years, oodles of other people and organizations have faced identical dilemmas.

For decades, employees frustrated with their employers' internal systems, applications, and devices could either lobby for internal change, grin and bear it, or quit. Starting in the mid-2000s, however, a new—if unsanctioned—option began growing in popularity.

Shadow IT

It's time to introduce another possibly foreign concept: shadow IT. It refers to the workplace technologies employees use that centralized IT departments haven't sanctioned.

A little background will put shadow IT into context.

Rewind to 1996 for a moment. Mary is a director of sales at a large retail outfit. She hates her employer's homegrown customer relationship management (CRM) system. Mary yearns for the days of a more contemporary alternative and pleads her case with the C-suite. (Yes, tech envy is a thing.) For the time being, however, there's not much she can do.

For years now, Mary has been able to quickly spin up her own CRM system and fly below the radar. Let's say that she signs her department up for Salesforce. Thank the advent of smartphones, cloud computing, and software as a service (SaaS). Small monthly charges are easier to sneak past internal auditors than massive one-time bills.

Mary may be happy with her new CRM system, but at least a few of her colleagues won't be when they find out. Nothing against Salesforce but, generally speaking, shadow IT can raise the eyebrows of security and compliance peeps. At a minimum, it sets a dangerous precedent: Employees can do whatever they want.

An itchy executive going rogue and signing up for Salesforce represents just one example of shadow IT. Others include using a personal email account to conduct company business or working on your own tablet in the office.

On the whole, shadow IT is far more common than one might think. In 2017, the Everest Group found that half of all enterprise purchases fell under the umbrella of shadow IT.[12]

Employees who deliberately opt to circumvent IT—or at least try—will usually tell you they have no choice. In an ideal world, they'd use sanctioned tools, systems, and applications. Sadly, those internal technologies are deficient and, in many cases, have been for a while. The rank-and-file frequently justifies going rogue because the ends justify the means.

This statement goes double for executives with profit-and-loss (P&L) responsibilities and lucrative stock and bonus packages. Antiquated tech or systems directly affect their compensation, making the costs of inaction just too great. Even *if* they get caught, they'll deal with the consequences later. Ask forgiveness, not permission.

At the risk of excusing potentially dangerous behavior that jeopardizes the entire organization, imagine if IT concedes the point. The CIO admits that the company's current technologies are deficient—or even a sad state of affairs. What if IT promises to make changes, upgrade systems, and introduce new tools as soon as possible?

In many cases, the CIO's assurances wouldn't placate anxious constituents. Those much-needed enhancements may take too long to arrive. IT often can't deliver the goods quickly enough. Although the reasons vary, an inability to hire more developers is one of the usual suspects.

The Interminable War for Tech and Data Talent

Before COVID-19, the Linux Academy estimated that two in three employers couldn't find qualified candidates to fill their open IT positions.[13] In 2017, Forrester Research predicted that the software developer deficit in the US alone would reach 500,000 by 2024.[14] (I'll bet my house that the pandemic makes that estimate look paltry by comparison.) IDC reported in its September 2021 Market Perspective that the global shortage of full-time developers will increase from 1.4 million in 2021 to 4.0 million in 2025.[15]

In its September 2021 report "The Tech Talent War Is Global, Cross-Industry, and a Matter of Survival," the venerable consulting firm Bain & Company revealed the intensity of the battle.[16] Across the board, the demand for both tech and data workers far exceeds their supplies. What's more, society doesn't just mint new programmers and data scientists overnight. As with doctors and lawyers, it takes time. Figure 1.3 shows how the demand for different tech-related jobs has mushroomed.

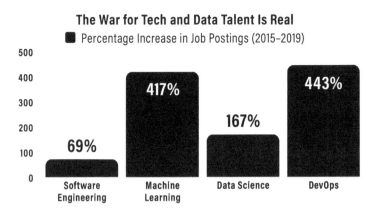

Figure 1.3: The War for Tech and Data Talent Is Real, Source: Bain & Co.

Note Bain's & Company's dates in Figure 1.3. They stem from prepandemic times. As a result, they fail to reflect the new business applications that the rise of remote work has heaped upon IT departments. That caveat aside, however, Bain's management recommendations hold up today, especially "find[ing] creative ways to widen their funnel of candidates."

Against this backdrop and that of #BlackLivesMatter, in February 2022, Google launched a $100 million Career Certificates Fund. From the Social Finance website, a national impact finance and advisory nonprofit:

> Supported by inaugural training providers Merit America and Year Up, learners upskill and earn Google Career Certificates—industry-recognized credentials that equip people with the skills needed to enter in-demand fields such as data analytics, IT support, project management, and user experience design. The program focuses on helping people from underserved communities access well-paying, high-growth jobs.[17]

The fund comes on the heels of the company's highly touted 2020 creation of a six-month career certificate program aimed at disrupting higher education. Google aspires to create equivalents to traditional four-year degrees for a fraction of the cost,[18] and early results have been impressive. More than 250,000 people entered the IT certificate program in its first two years.[19]

Hiring part-timers is another potentially promising way to increase the pool of technical candidates. Unfortunately, that dog won't hunt, says Arnal Dayaratna, a VP of research at the global market intelligence firm IDC:

> While part-time developers provide an invaluable resource for organizations to continue digitizing business operations and processes, there is no substitute for trained full-time developers that have the skills to architect digital solutions with due consideration for their long-term viability, scalability, and security.[20]

Jargon aside, Dayaratna is right. It's a half measure.

And the severe talent scarcity isn't confined to software developers and general tech workers. Data wizards are also in short supply. Consulting firm QuantHub in 2020 reported a shortage of 250,000 data scientists with no end in sight.[21]

As further proof of the imbalance and the need to address it quickly, consider higher education. For the past six years, colleges and universities have been scrambling to create degree and non-degree programs for data analysts and scientists.[22] I should know. I used to work for one of them.

Retention Issues and Remote Work

Forget hiring *new* developers. Thanks to the pandemic, employers as a lot are struggling to retain their *current* ones.

Consider the words of Marko Kaasila, SVP of product management at the Qt Company. After his firm copublished research with Forrester in June 2021, he spoke with Owen Hughes of TechRepublic:

> The unforeseen need for rapid digital transformation in recent months has placed a huge drain on developers who have not been equipped with the tools they need to manage the dramatic rate of change. The welfare of software developers in today's fast-paced world has been overlooked as companies digitally innovate in order to survive.[23]

The acceptance of remote work adds pressure to IT execs who want their employees to, you know, show up at the office occasionally. In April 2020, researchers Jonathan Dingel and Brent Neiman of the University of Chicago approximated that 100 percent of US technology workers can do their jobs remotely.[24]

Think about that number. Every. Single. One.

As a group, technology workers aren't too keen on the idea of returning to an in-person, Monday–Friday, 9–5 work schedule. In May 2022, Morning Consult surveyed 638 hybrid or fully remote techies. Roughly 60 percent of respondents said they weren't interested in returning to full-time, in-person work. Good luck calling their bluffs. In the current labor market, they're holding pocket aces. They won't be unemployed for long—and they know it.

Organizations face an uphill battle in finding and retaining talented tech workers. Forcing these employees to return to a prepandemic work environment and schedule only intensifies the challenge. And remote work adds yet another fly in the ointment. From "Microsoft Digital Defense Report":

> While most industries made the shift to remote work due to the pandemic, it created new attack surfaces for cybercriminals to take advantage of, such as home devices being used for business purposes.[25]

Employees who work from home and connect to their home networks pose a new array of security risks[26] outside the scope of this book. Suffice it to say that securing a remote or hybrid workforce requires new software, employee training, and possibly hardware. Supply chain and geopolitical issues further complicate the latter. As a result, IT departments are shifting dollars away from application development—the *Dev* part of DevOps.

A Way Forward

This chapter has shown that IT departments face an untenable status quo. Their lack of resources prevents them from meeting their constituents' growing needs. The current labor market makes it nearly impossible for them to procure those resources. Faced with an often-unresponsive IT department, shadow IT is proliferating, giving many executives agita.

At this point, you might be thinking, "Don't bring me problems. Bring me solutions."

It's among the most hackneyed management tropes—and has been for years. Maybe your boss has uttered those words at some point, and you rolled your eyes.

IT departments routinely unable to meet their constituents' needs can choose one of three options.

#1: Ask Them to Be Patient

"We know that you need critical apps to meet your business goals. They're coming—eventually. Really. We promise. We just don't know when."

Good luck with that.

#2: Encourage Them to Go to Development Shops

"Fine, we admit it. We can't deliver the goods. Go ahead and find other developers who can."

Outsourcing app development can be a particularly tough needle to thread. Horror stories abound, and the data supports this contention. Consider the words of Tom Dunlap, director of research for Computer Economics:

> Application development can be a tricky outsourcing category to get right. Application developers are expensive and, in many cases, are rightly seen as options to outsource. But our moderate service-satisfaction numbers show there is a risk associated with this type of outsourcing.[27]

Although the second alternative is better, it's hardly ideal.

#3: Allow Nontechnical Employees to Develop Their Own Business Apps

The third option is hands down the most plausible one. Not coincidentally, it's also the subject of this book. Together, an emerging group of tools (called *low-code/no-code*) and tech-savvy employees (called *citizen developers*) are democratizing the development of business applications. Chapter 5 explores the group in depth.

Before continuing, however, we need to cover how firms have historically built and deployed new tech. Let that serve as the starting point for the next chapter.

Chapter Summary

> In case any of us forgot, workplace tech keeps the wheels moving. The pandemic gave us an essential reminder of its primary import.
> The marked shortage of developers means that IT departments can't meet their constituents' needs. They're unable to develop and deploy critical business applications and systems.
> This challenge is neither small nor ephemeral.
> Few tech workers want to return to the office on a full-time basis. This reality makes finding and retaining them even more problematic.

Chapter 2

A Comically Brief Overview of Conventional Business Technology

"Hindsight bias makes surprises vanish."

—DANIEL KAHNEMAN

What applications and systems does your employer run? What's the history? How did it develop and deploy them?

I find these questions endlessly intriguing, but what about you?

Depending on your age, role, background, and tenure, these questions may not concern you one bit. Maybe you're even blissfully unaware.

After all, to the layperson, what's the difference? Garden-variety accounts payable clerks, security guards, finance analysts, and administrative assistants don't need to know if their employer runs the latest version of an application. Does System X rely on a mainframe computer, client-server architecture, or cloud computing?

At any given workplace, relatively few people need to know the underpinnings of each technology.

I'll concede the point. For our purposes, the intricacies of every type of business application and system aren't terribly important. (Even many seasoned IT professionals freely admit that these critical areas aren't all that interesting.) We do, however, briefly need to discuss two tech topics:

> The types of systems and applications that organizations use.
> How organizations have historically deployed them.

My rationale here is straightforward and threefold.

First, business applications and systems have never existed in a vacuum. At least, they shouldn't. Although their types and foundations could be a black box to you, each one relates to—and typically interacts with—other systems and applications. Second, to some extent, workplace tech affects every employee's performance, as well as their interactions and overall satisfaction. The same truism applies to their teams, departments, divisions, and employers as a whole.

Finally, and as Part II manifests, workplace tech is morphing before our eyes. Software development is no longer the exclusive purview of proper programmers. Millions of nondevelopers are crashing the party—and they would benefit from knowing some basics before crashing it.

Let's go.

Defining Our Terms

Using a business application to create a relatively sophisticated document, spreadsheet, or presentation primarily or exclusively for yourself is one thing. We've all done it at some point in our careers—in some cases, thousands of times. Designing and building an application or system that others will understand and consistently use, however, is something else entirely.

Admittedly, the difference between a business *application* and a *system* has always been fuzzy. It's not as if some governing body decides which is which. As Figure 2.1 demonstrates, sometimes the two entities overlap.

Traditional Applications and Systems

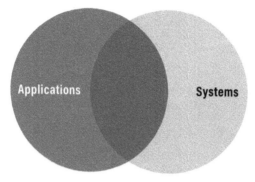

Figure 2.1: Traditional Applications and Systems

The answer to the application-vs.-system question may depend on how the person, team, department, or entire firm uses the creation.

As a rudimentary example, consider Google Sheets. Initially launched in March 2006,[1] the Microsoft Excel knockoff formerly fell under the G Suite umbrella. In October 2020, the company rebranded its productivity suite as Google Workspace.[2] (Note to Google CEO Sundar Pichai: Please stop mucking with the names of your company's products and services. You're confusing the hell out of us.)

Names aside, more than 6 million organizations pay to use Sheets to track different types of information.[3] (Before COVID-19, that number was 5 million.[4]) Millions more rely on the free version.

Unlike the Excel workbook that lives on your computer's hard drive,[*] however, Sheets lives on the web. It allows employees in multiple departments and on teams to concurrently enter, view, and update information on their mobile devices. Sheets supports concurrent users. In this way, it can serve as a crude system of sorts. Still, no one would confuse it with the proper business systems described in this chapter.

Note that the overlap between the circles in Figure 2.1 will only increase. As we'll see throughout this book, low-code/no-code has advanced. Today, no-coders and low-coders build valuable applications and lightweight systems that others can easily access and use no matter where they are.

Popular Types of Business Applications

With that critical distinction out of the way, Table 2.1 presents some of the most popular types and examples of modern business applications.

Category	Examples
Productivity	Word processors, spreadsheets, and the like. Think Microsoft Office 365 or Google Workspace. Many open-source zealots embrace OpenOffice.
Professional content creation and design	Adobe Creative Cloud is the big kahuna for creative types. AutoCAD lets architects, landscapers, interior designers, and real estate developers design and draft multidimensional blueprints, schematics, and models. Canva and Figma allow for increasingly collaborative design over the web.
Communications	Email, internal collaboration hubs such as Slack and Microsoft Teams, and older videoconferencing tools à la Skype and Cisco's Webex.

---

[*] Microsoft has added the ability to share Excel workbooks over the web.

Category	Examples
Security	Subcategories include antivirus, firewall, network security, virtual private networks (VPNs), and employee surveillance.
Data visualization and analysis	Tableau and Microsoft Power BI are two powerful tools that let nontechnical users analyze data and create interactive data visualizations.
Statistical analysis	IBM's SPSS and the SAS suite allow for sophisticated statistical analysis, modeling, and much more.

Table 2.1: Popular Categories of Business Applications

In all but the craziest scenarios, all these are applications; they're not proper systems.

Popular Types of Business Systems

In the interests of brevity and staying true to this chapter's title, I'll lump all business systems into three types.

Back-Office Systems

Enterprise resource planning (ERP) systems represent one major category. ERP systems from Oracle, SAP, Workday, Microsoft, NetSuite, and others help tens of thousands of organizations manage day-to-day business activities. You might also hear people refer to them as *back-office systems* because the employees who use them don't directly interact with a firm's customers. Employees in the following departments rely upon ERP systems to do their jobs: human resources, payroll, accounting, procurement, and supply chain management.

Front-Office Systems

Next up is another central bucket: CRM. These systems are equally essential to an organization's operations but in a different way. Employees who use them interact directly with customers and prospects. Software from Salesforce, Oracle, Microsoft, and others allows firms to manage all their relationships and interactions with existing and potential customers in a single place. Ideally, CRM systems encompass all of a firm's marketing, sales, and customer service efforts. (In practice, however, this is rarely the case.)

Think of CRM and ERP systems as yin and yang. The former handles essential front-office functions, while the latter steadies the back-office ship. (In truth, the line between front- and back-office systems has started to blur over the past decade, but we don't have to dive too deep into that subject here.)

Here's the main point about commercial-off-the-shelf (COTS) software: Whether it's an affordable accounting package like QuickBooks or a full-fledged enterprise system like Microsoft Dynamics 365, the organization purchases existing software and configures it. The system ships with functionality to enroll employees in benefit plans, pay vendors, and the like. Companies merely need to set up the system, train people how to use it, load data, test it, and go live. (Sounds simple but, as I write in *Why New Systems Fail*, this process is riddled with problems.)

Everything Else

The final bucket includes the following subcategories represented in Table 2.2.

Type of System	Description
Product lifecycle management (PLM)	PLM systems allow firms to manage a product's "life." I'm referring to its birth, development, sales, and death. SAP and Oracle are two popular choices.
Learning management (LMS)	Just about every college professor uses a learning management system, although sometimes it's begrudgingly. Blackboard, Canva, and Google Classroom are just a few examples.
Ticket management	Although there are others, Zendesk and Freshdesk immediately come to mind.
Content management	HubSpot and WordPress are common choices to power organizations' websites. Note that large enterprises typically require industrial-strength hosting services. For example, Dropbox and National Geographic don't pay GoDaddy $8/month; they pony up far more to WPEngine.
Fraud detection	Credit card companies and banks rely on intricate systems and machine learning to spot bogus charges and criminal activity. I don't know what Chase uses for this purpose, but I can guarantee that it's a hell of a lot more sophisticated than Microsoft Access or Airtable.
Healthcare	Hospitals, doctors, and dentists run a bevy of systems to perform vital tasks. They include tracking patient medical records, managing their practices, monitoring supplies and inventory, and more. McKesson and Epic are two of the big players.
Knowledge management	Atlassian Confluence, Document360, and other souped-up wikis allow organizations to capture and store invaluable institutional knowledge. Employees can quickly locate internal documents, answers, resources, and much more—at least in theory.

Table 2.2: Additional Categories of Business Systems

To state the obvious, this list isn't remotely comprehensive. The main point is that organizations need quite a few different types of systems to keep the lights on. In many cases, there's space between

and among them to create glue systems and applications. Low-code/
no-code (LC/NC) tools excel in this area.

How Organizations Launch New Workplace Technologies

At a high level, organizations have historically developed and
deployed new applications and systems in several ways. Here's a
brief primer on each.

Build From Scratch

People of a certain age remember the early days of the web. You
know, back when Amazon only sold books online. On *Today* in
1994, Katie Couric and Bryant Gumbel typified the general public's
reaction to this weird, burgeoning technology.[5]

As someone with a front-row seat, I've witnessed how the web
and ecommerce have evolved. There's a reason that Brad Stone's
bestselling book on Amazon is titled *The Everything Store.* Nearly
three decades after that infamous NBC segment aired, there are
dozens of ways to remit money to others online. Believe me: It
wasn't always so easy.

In late 1998, entrepreneurs Peter Thiel and Max Levchin launched
a venture to achieve that very goal. Ultimately, Confinity created
one of the first ways for businesses and individuals to securely send
and receive electronic payments over the web.[6] Quickly rebranded
as PayPal, the company needed to build custom software from
scratch because it simply didn't exist.

These types of greenfield development projects are sometimes
not only wise but necessary. Exhibit B: In 1994, no one could buy
an auction program off the shelf. For eBay to go from crazy idea to
reality, Pierre Omidyar and his team would have to write code that
allowed people to simulate a real-world auction and bid on items

in real time. They did, and the rest is history. (Unsolicited book recommendation: *The Perfect Store: Inside eBay.*)

These cases aside, though, the idea of building custom software is often misguided or just plain dumb. Case in point: In the early 1990s, an international pharmaceutical giant built its own janky HR systems. (By that rationale, Microsoft should manufacture its own aspirin.) When superior ERP systems hit the market, management dillydallied for years. The company eventually purchased one but deployed it in a half-assed manner. After wasting millions of dollars, it finally conceded defeat, ripped off the Band-Aid, and implemented a new ERP system the right way.

General Advantages

Organizations that create their own custom software can benefit in the following ways:

> **Meet a unique business need.** PayPal and eBay serve as instructive examples.
> **Use the better mousetrap to launch the company.** Google did this in 1998. Its search engine wasn't 10 percent better than existing ones; it was *10 times* better.
> **Sell that software to other organizations.** Amazon Web Services started as an internal project before the company decided to license the service to other firms. As another example, company founder and current executive chairman Jeff Bezos also owns the *Washington Post*. The *Post* now licenses its proprietary content management system to other organizations. The CMS represents its most valuable asset.[7]
> **Avoid relying upon the wares of a competitor or foreign entity.** Based on the success of AWS, Microsoft went all-in

with cloud computing under new CEO Satya Nadella. Microsoft Azure today represents a viable option to AWS.

> **Technical reasons.** Examples here vary. For instance, Dropbox abandoned AWS and built its own data- and file-storing infrastructure so the company could maintain greater control over its performance.[8]
> **Save money.** Compared to the alternatives discussed in this chapter, sometimes it's cheaper to build and own a proprietary system.

Potential Drawbacks

Although this isn't a comprehensive list, some of the main disadvantages of this approach include the following:

> **Effort:** Building unique enterprise systems is expensive, time-consuming, and prone to errors.
> **Talent deficiency:** The organization lacks the right talent to build the system it thinks it needs. As we know from the previous chapter, hiring and retaining software developers in the current environment isn't easy.
> **Lack of an existing user community:** As a result, you won't be able to benefit from others' experiences and expertise. Oh, and no one else will find its bugs for you.
> **Misguided need:** Existing market solutions are sufficient, and the business case for reinventing the wheel is weak. I've never met anyone who stays at an employer solely because its internal systems are excellent. Think of the latter as a hygiene factor: Its presence guarantees nothing, but its *absence* poses problems.

Buy and Configure Existing Software

Building systems from scratch is far from the only option available to companies needing new tech. Not only is it not even the best option, in many cases it represents the worst possible decision.

It's time to talk about commercial-off-the-shelf software—a software engineering term for ready-made products. Note that the "shelf" in question could be physical or digital. In the case of the former, individuals and small business owners frequently purchased software from big-box retailers like Circuit City and CompUSA. Installation consisted of floppy disks and then compact discs. The CIO of Home Depot never walked the aisles of Circuit City searching for an enterprise-grade relational database management system.

General Advantages

Potential benefits of buying and configuring COTS software include increased reliability and reduced total cost of ownership. Building a custom CRM or ERP from scratch rarely makes sense; COTS almost always represents the better use of company funds.

Configuring an existing, feature-rich system is at least an order of magnitude easier than starting with a blank screen and zero functionality. Security is also a plus. If a customer discovers a bug in a system or application, the vendor will issue a patch.

Potential Drawbacks

Depending on the scope of the endeavor, expect significant capital expenditures from the start. Someone will need to write a big check to purchase the system and kick off the project implementation. And the costs won't end there. Software vendors typically charge their clients 20 percent per year to maintain the system.

Unlike the bespoke systems that firms construct, COTS won't meet every possible business need. When an organization replaces its homegrown system with Oracle, SAP, or something else, the new systems will invariably differ from their predecessors. Employees frequently bristle at changes to their routines and tools. Customizing the COTS system is an option, but that can cause massive problems and invalidate vendor support.

Finally, the software vendor may decide to retire—or *sunset*, in the parlance of the industry—key modules or applications of the system. As a result, the organization has to either upgrade or find independent support if it wants to keep using the vendor's legacy tech.

Rent From a Software Vendor

In March 1999, four men began working in a tiny one-bedroom apartment in the Telegraph Hill neighborhood in San Francisco.[9] Before they knew it, Parker Harris, Frank Dominguez, Dave Moellenhoff, and Marc Benioff had changed how millions of organizations purchased and deployed software. (No, I'm not exaggerating.) If there's a Mount Rushmore of tech, Salesforce is on it.

We take it for granted today, but not *that* long ago, plenty of tech luminaries considered the idea of renting software absurd, impossible, or verboten. In 2008, Oracle cofounder and then-CEO Larry Ellison infamously lambasted cloud computing as "idiocy."[10]

Whoops.

The rise of cloud computing in the 2010s changed the minds of its most hardened skeptics, Ellison included. Benioff and company birthed a novel idea: software as a service. It meant renting systems and applications in lieu of purchasing them.

In the years that followed, countless startups have adopted the SaaS business model out of the gate or, in the case of behemoths such as Microsoft, eventually switched to it.* Why sell a person or company software every five to seven years? Why not bill them every month in perpetuity? (For more on this subject, see Nicholas Carr's excellent book *The Big Switch: Rewiring the World, from Edison to Google.*)

General Advantages

The major benefits of SaaS include:

> **Lower up-front costs:** Think moderate monthly operational expenses, not massive one-time capital expenditures. There's no need to purchase servers and other hardware; the vendor takes care of everything for you.

> **Nonexistent maintenance costs:** Companies no longer host their own systems. The *on-premises* model requires employing pricey IT staff to manage, maintain, and upgrade systems. There's no need to do that in the SaaS universe because the software vendor takes care of that. (Remember from Chapter 1 that the demand for these folks is white hot.)

> **Guaranteed reliability:** Depending on the specific plan selected, end user license agreements—aka, *software license agreements*—generally offer uptime guarantees of 99.99 percent. Expect refunds or future credits if a vendor fails to hit that number due to a system or network outage.

> **Accessibility and device agnosticism:** Software vendors build modern SaaS apps and systems to run on just about anything, anywhere. There's no need to head to the office.

---

* To be fair, plenty of customers from software vendors have been none too pleased about paying monthly fees after so many years of the old business model.

> **Scalability:** Organizations can easily accommodate changing business needs and budgets. Growing firms can quickly add users, compute power, and other resources as needed; shrinking ones can cut back to reflect their new realities. In either case, SaaS minimizes waste and shelfware (the practice of paying for unused software).

Potential Drawbacks

The major handicaps of SaaS—and, by extension, cloud computing—include:

> **Loss of control:** Many old-school companies and IT execs refuse to allow a third party to run critical aspects of their firms' infrastructure.

> **Complications with legacy systems and tech:** Integrating older systems and applications with certain types of clouds can be difficult and expensive. The idea that someone can click a button that magically ports "everything over to the cloud" is preposterous.

> **Throttling due to noisy neighbors:** Say that your mobile carrier is AT&T or Verizon. In October, you pass your allotted threshold for monthly data usage. Once this happens, your carrier intentionally slows down your browsing speed—a practice known as *throttling*. Similarly, assume that your organization hosts its infrastructure on a public cloud with other companies. Your neighbors' activity may well adversely affect the performance of *your* system and applications. (For much more on this topic, Google "multitenancy" and "rate-limiting.")

> **Higher *potential* overall tech costs:** Relative to independently owning and operating a system, it's possible for an organization to incur *higher* costs by paying on a per-user basis. For example, $15 per user per month may represent a rounding error for a five-person law firm if its attorneys charge $600 per hour. To a 25,000-employee organization, however, that rate may become untenable. (To be fair, SaaS vendors understand this arithmetic all too well. As a result, they tend to offer discounted enterprise, all-you-can-eat plans beyond a certain user threshold.)

> **Limited customizability:** Let's say you rent an apartment on a two-year lease. You won't be able to remodel the kitchen or redo the master bathroom. The same principle holds with renting software. Most SaaS systems and applications limit their clients' ability to customize the software, especially compared to the building and buying options discussed earlier in this chapter.

Discerning readers will note that I'm intentionally ignoring any purported security concerns. Yes, hacks happen, but the idea that on-premises apps and systems are fundamentally more secure than all cloud-based ones is risible.[*]

Adopt Another Approach

This chapter has covered a great deal of terrain by design. I'll wrap up by briefly mentioning open-source software and hybrid approaches.

---

[*] For more basics on cloud computing, see https://tinyurl.com/philcloud.

Embrace Open-Source Systems

If you're not enamored with Salesforce and NetSuite, a "free" solution like Odoo might be just the ticket.

An open-source alternative exists for just about every conceivable commercial business system or application. Yes, these programs are free to download and use under one license or another. As the tried-and-true saying goes, think free speech, not free beer. "The community" isn't going to configure and test your instance of osTicket, a popular open-source ticketing system. Unless you adopt a DIY approach, you'll have to pay others to customize the system, implement it, maintain it, and eventually upgrade it.

Hybrid

Plenty of organizations mix and match. That is, they select, implement, and maintain different types of systems based on business needs. Perhaps Company X hosts its own legacy ERP system but adopted Salesforce as its CRM two years ago. Maybe an essential reporting or tracking system is so old that moving it to "the cloud" isn't feasible.

Hybrid approaches are the norm after a merger or acquisition takes place. After the deal closes, firms need to plan how to migrate the acquired company's systems to those of the acquirer.

Development Methods

The decision to buy, rent, or build is a biggie—and it's not the last decision that a firm or department has to make. At a high level, firms can deploy new applications and systems via each of the following methods:

> **Waterfall:** The Phase-Gate approach is sequential, rigid, and based on being able to predict the future. The product arrives at the end.
> **Agile:** Scrum, Extreme Development, and other flexible approaches deliver the system in increments.

Table 2.3 provides a short, high-level summary of these two methods across several attributes.

Characteristic	Waterfall Approach	Agile Approaches
Approach	Sequential, fixed	Incremental, continuous
Team Attributes	Large, specialized	Small, cross-functional
Mindset	"We can control everything."	"We can't control anything. Let's not even try."
Planning	Long-term	Short-term
Building Blocks	Fixed business requirements	Flexible, constantly evolving user stories
Estimates	Absolute. "It will take 3.25 hours to mow Marie's lawn."	Relative. "I'm not sure how long it will take to mow Marie's lawn, but hers is bigger than Walter's. It will take more time."
Feedback	Takes place once when the team has completed the project	Occurs routinely at the end of each sprint
Loop Type	Open	Closed
Testing	At the end of the development phase, in a single phase	At the end of every sprint
Batch size and delivery	Large, delivered at the end of the project	Small, delivered regularly

Table 2.3: Comparing the Mainstream Software Deployment Approaches

I'm specifically ignoring Agilefall, Waglile, and other bastardized amalgams. You can't get a little bit pregnant.

Part I has covered a great deal of ground, but there's no getting around it. By and large, IT departments simply can't fulfill their constituents' insatiable demand for new business applications and systems—and enhancements to existing ones. For this reason, a robust new set of development tools has exploded in popularity—enabling nontechies to become *de facto* software developers. Let that serve as the starting point for Part II of this book.

Chapter Summary

> Organizations have historically developed and deployed new systems and applications in five ways.
> Each alternative inheres different costs and benefits.
> Some firms mix and match; that is, they'll rent specific applications or systems but host others themselves.
> At a high level, organizations can develop and deploy software in a rigid, sequential manner or in a piecemeal fashion.

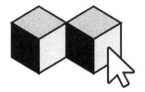

< Part II>

Reimagining Application Development

Chapter 3

Why Low-Code/No-Code Changes Everything

"Their MO is that they're good."

—AL PACINO AS LIEUTENANT VINCENT HANNA, *HEAT*

Most new business applications and systems have traditionally come from experts trained to build them. Whether they worked for software vendors, development shops, or internal IT departments was moot. Experienced developers rolled up their sleeves and hammered away at their keyboards for days, weeks, months, or even years.

But what about most *future* business applications and systems?

A massive shift is already taking place. An enthusiastic group of decidedly nontechnical employees is taking over most of the development reins. Their primary instrument isn't a keyboard; it's a computer mouse.[1]

Part II of this book answers the simple yet profound question: How is this happening?

The short answer is twofold. First, nontechies are using an increasingly powerful set of tools to build all sorts of valuable business applications. Yes, I'm talking about low-code/no-code, the focus of this chapter. Chapter 5 checks the second box and properly introduces this group as citizen developers.

Let's begin with a brief but necessary history lesson.

Low-Code/No-Code: Building Blocks and Precursors

The tools covered in this book didn't just suddenly arrive; they mostly evolved from fourth-generation programming languages. Depending on your age, interests, and background, the previous sentence may have prompted you to respond with the highly technical question, *Huh?*

Previous Generations of Programming Languages

I'm not qualified to write a detailed opus about the history of computers or programming. Work in the tech world long enough, though, and you pick up a few things.

Like all contemporary software, current LC/NC tools descend directly from previous programming languages. To this limited extent, a brief overview of the topic makes sense. Table 3.1 summarizes the different generations of mainstream coding languages.

Generation	Level	Brief Description and Notes	Specific Language(s)
First (1GL)	Low	This refers to machine-specific, nonportable instructions that a particular computer can interpret sans further translation.	N/A
Second (2GL)	Low	A utility program or assembler converts language into executable code that the computer can understand.	Assembly

Generation	Level	Brief Description and Notes	Specific Language(s)
Third (3GL)	High	Portable languages that let developers write programs that work more or less the same on a specific type of computer. Unlike previous generations, users could take the code with them and make minor tweaks; it wasn't wedded to one particular machine.	C, FORTRAN, COBOL, Pascal, C++, Java, and BASIC
Fourth (4GL)	High	Languages in this generation rely on integrated development environments (IDEs). IDEs package a proper code editor, compiler, debugger, and graphical user interface. A moderately technical layperson could look at the code and understand what's happening. 4GL serves as the roots for modern-day low-code/no-code tools.	Perl, PHP, Python, Ruby, and SQL
Fifth (5GL)	High	Predicated on solving problems using constraints given to the program, not specific lines of code. (Yes, we're talking about artificial intelligence and algorithms here.)	OPS5 and Mercury; note that some libraries extend the power of 4GLs to perform similar tasks

Table 3.1: The Generations of Programming Languages

Evolution > Revolution

Table 3.1 helps us view the current low-code/no-code movement in its proper historical context. Think of LC/NC as more of an evolution than a revolution.

Dave Farley is a longtime independent software developer and a prolific author. Speaking about LC/NC in February 2022, he correctly notes:

> There's nothing new to this idea, really. In some ways, this describes the evolution of programming in general. What were the first compiled languages, if not attempts to make programming easier than dealing with machine code?[2]

Farley goes on to wager that machine programmers would have derisively labeled people writing COBOL statements as low-coders back in the day. I suspect that he's right. Haters gonna hate, right?

What's more, whether we realize it or not, we're all still regularly using applications and systems written in the older programming languages described in Table 3.1 for all sorts of purposes. Put differently, reports of the death of systems based on, er, older programming languages are greatly exaggerated—a point that the pandemic accentuated. As Owen Hughes writes for TechRepublic:

> In April 2020, hundreds of thousands of residents submitted applications to the State of New Jersey's unemployment system, leading to a 1,600 percent increase in claims which quickly overwhelmed its COBOL-based mainframe.[3]

More relevant for our purposes, however, is another example: applications that let nondevelopers build their own applications. (That's right. I just got all meta.)

Visual Programming

Although they went under a different moniker, drag-and-drop development applications have been with us for more than a quarter-century. Tools that have fallen under the *visual programming* umbrella have let users build their own apps despite lacking any proper coding knowledge. (For an example of contemporary visual programming, check out Blockly.[4])

It's hard to overstate the seismic nature of this shift. Laypeople generally find the mouse to be more accessible than the keyboard. Visual programming lets them perform functions by graphically manipulating elements rather than by specifying them via text. The shift in primary instruments has encouraged budding programmers in areas such as education, automation, video games, music, and animation. (Cue Steve Jobs's quote about the computer being a bicycle for the mind.)

Since this is a book about workplace technology, however, let's focus on popular visual-programming applications that business folks have used over the past thirty years. I'll present them in chronological order of when they arrived.

Microsoft Access

Industrial-grade databases harken back to the 1960s. (Even today, the enterprise systems discussed in Chapter 2 couldn't exist without them.) Although incredibly useful for managing vast amounts of data, these databases typically represented overkill for most businesses, never mind small teams and individual employees. Against this background, *personal* databases began to gain steam in the late 1970s and early 1980s. Examples included Paradox, dBase, and FoxPro.

Realizing that the trend would only intensify, Microsoft hopped on this bandwagon. The company initially released Microsoft Access in November 1992 at the COMDEX convention in Las Vegas.[5] Critically, Access shipped with a graphical user interface (GUI). As a result, by merely dragging and dropping, its users could create reasonably sophisticated databases, forms, queries, and reports.

Knowledge of SQL didn't hurt, but effectively using Access didn't require it.

In the ensuing three decades, Access has proven remarkably versatile, valuable, and, unsurprisingly, difficult to retire. Its users run the gamut. Small and large mature organizations in just about every industry have hopped aboard the Access train and don't want to get off—and not just in the US. As a result, some industry insiders have termed it "the software that refuses to die."[6]

Microsoft FrontPage

In January 1996, Microsoft acquired Vermeer Technologies and its crowned jewel, FrontPage, a what-you-see-is-what-you-get (WYSIWYG) HTML[*] editor and website administration tool. Microsoft included FrontPage in its ubiquitous Office suite from 1997 to 2003. Figure 3.1 displays a screenshot of what was—believe it or not—cutting-edge software at the time.

Microsoft FrontPage User Interface

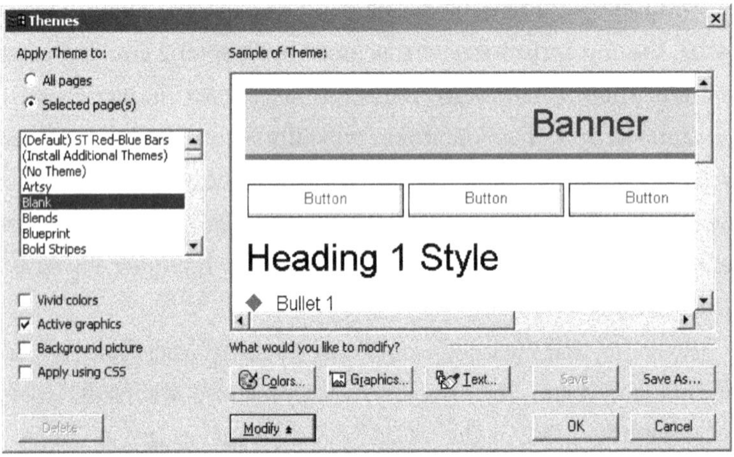

Figure 3.1: Microsoft FrontPage User Interface, Source: Microsoft

---

[*] Short for *Hypertext Markup Language,* an underpinning of the web to this day.

Unlike its Microsoft sibling Access, however, FrontPage lacked staying power. A few years after the acquisition, it ceased to be compatible with most web servers at the time.* In 2003, Microsoft euthanized FrontPage and encouraged its customers to use its Expression Web and SharePoint Designer products.

Dreamweaver

Originally launched in December 1997 by Macromedia, Dreamweaver is a WYSIWYG website builder and code editor that's still around today. Adobe Systems acquired Macromedia in December 2005, and some would argue that the move represented the beginning of the end of Dreamweaver.

On Quora and Reddit, you'll find plenty of spirited debates over the reasons for Dreamweaver's decline. I don't want to start a holy war. Suffice it to say that it's not nearly as popular as it used to be. At present, only 0.3 percent of all websites rely on it.[7]

On a personal level, I taught ten sections of the information system design course as a college professor. Of the fifty projects that I supervised from 2016 to 2018, more than half involved building websites. A grand total of zero projects involved Dreamweaver; most involved the LC/NC tools Squarespace and WordPress.

Squarespace

Anthony Casalena started the blog hosting service Squarespace from his dorm room at the University of Maryland in April 2003. Eventually, Squarespace swapped out its coding back end with an exclusively drag-and-drop user interface that relied on the no-code design system Fluid Engine.[8] As of this writing, the

---

* For a more comprehensive history on the topic, see https://tinyurl. com/ps-FP2023.

content management system is about seven times more popular than Dreamweaver.[9]

WordPress

WordPress debuted on May 27, 2003, around the same time that Microsoft was putting FrontPage out to pasture. Although originally conceived as simple, free blogging software, WordPress quickly evolved into a robust content management system that millions of laypersons and businesses use in creative ways all these years later. As of this writing, WordPress currently runs more than 43 percent of all websites.[10]

Bubble

Started in 2011, Bubble lets anyone with a mouse build a variety of powerful apps. Designing and deploying modern, interactive apps is remarkably simple with Bubble. I'll let you peruse its website yourself,[11] but its feature set is impressive. (Chapter 7 contains an in-depth case study of Bubble in action.) In October 2021, the company raised $100 million in Series A funding.[12]

Coding for the Masses

It's instructive to view the six distinct pieces of software in the previous section as part of a longer-term trend. Writing for *Wired* in May 2020, Clive Thompson astutely notes:

> The emergence of no-code is, in a sense, the ur-pattern of software. We've been drifting this way for years. Websites at first were laboriously hand-coded until blogging CMSs automated it—and blogging exploded. Putting video online was a gnarly affair until YouTube rendered it frictionless—and vlogging exploded.[13]

The trend toward democratizing tech is anything but nascent. (As an aside, Thompson's 2019 book *Coders* is excellent.)

The technically inclined have been creating websites and standalone apps for decades. But what about general efficiency nuts? For their part, these power users have been making tweaks, automating manual tasks, and saving time within their current programs and on their computers for just as long. Here are a few examples:

> Macros for different Microsoft Office programs are an absolute godsend. Ditto for creating custom menus and tweaking existing ones. With respect to the former, the time saved from recording and using simple ones can be substantial. (If you can't access that functionality today, it's because of security concerns. Microsoft is disabling macros by default.[14])

> Mac users have been able to create custom apps since 2005 via Automator, a visual-scripting technology. Quit each open application one by one if you like, but a single mouse click can also get the job done.[15] Apple has announced that its Shortcuts app is replacing Automator for MacOS,[16] but custom automations aren't going away; they're just moving elsewhere.

> TextExpander and equivalent apps let users create custom shortcuts that replace text through recorded shortcuts.

All of this is to say that this trend isn't new, but the name of the movement is.

The Great LC/NC Rebranding

Google Trends tells us that the term *no-code development plat-form* first emerged in its search results in 2004.[17] Despite that brief appearance, no-code as a concept remained somewhat under the radar for a decade.

In 2014, Forrester Research released a report that ultimately resulted in the widespread use of its sibling: *low-code.* (It was hardly the first time that a consulting outfit has successfully coined a new technology or business term.) In "New Development Platforms Emerge for Customer-Facing Applications," Forrester nailed it. The company noted that "low-code platforms enable rapid delivery of business applications with a minimum of hand-coding and minimal upfront investment in setup, training, and deployment."[18]

Consider the words of Clay Richardson, one of the report's coauthors. Richardson describes how the origins of *low-code* evolved from a separate 2011 Forrester publication on the emergence of new tools that let everyday users build custom productivity applications:

> Around [that] time, Forrester began highlighting the need for workflow and process-change platforms to become more lightweight and nimble. We decided to bring these two streams of research together in 2013 to focus on speeding up development for customer-facing applications, which is where the term 'low-code' really stuck. Through dozens of interviews with companies, we found that the low-code imperative was strongest for teams trying to quickly build apps in response to customer demands and customer needs.[19]

Initially, Forrester grouped low-code tools into the following three categories that I've simplified a smidge:

> **Business process management (BPM):** In plain English, BPM allows organizations to optimize how they get things done. Paying vendors and paying employees are just two examples.
> **Content management:** This includes blog posts, articles, videos, audio, and other forms of, you know, content.
> **General purpose:** This is everything else that doesn't fall into the previous two categories.

It wasn't long before the old moniker *visual programming* started losing its luster. In its stead, industry types gradually started using the term *low-code* and its cousin *no-code*. Since then, LC/NC has come to classify a group of user-friendly, affordable, and powerful tools that let people create custom apps.*

The LC/NC tent is nothing if not inclusive: Nontechies, tried-and-true coders, and everyone in between can rapidly build robust business applications for their colleagues, partners, and customers.

This last point is critical: The ultimate output of an older visual-programming application was typically a standalone Microsoft Access database or souped-up Excel workbook with macros. It lived on a single employee's desktop. Most of the apps in question supported only a single user, lived on a single machine, or both. Deployment on a broader scale typically involved IT, Microsoft SharePoint, and intranets. Click-and-publish it was not.

The cohort of powerful LC/NC tools described in Chapter 4, however, opens a whole new world:

> Native support for concurrent users? Check.

---

* When the cool kids use the term LC/NC, they mean "low class/no class." At least that's what Urban Dictionary tells me. I'm not all that cool.

> The ability for people outside the organization to quickly use applications? Check.
> Easy, rapid deployment? Check.

Yep, it's a totally different beast. I'll return to the advantages of using low-code/no-code tools later in this chapter, but we've still got a little work to do.

Letters Matter: Distinguishing Between NC and LC Tools

At a high level, low-code/no-code allows nondevelopers to create applications despite lacking much—if any—knowledge of software development. (Chapter 5 properly introduces the *citizen developer*.) Before proceeding, a brief delineation between NC and LC makes sense.

No-Code

No-code tools forbid their users from adding additional code. Period. You'll be able to get away only with using your mouse. The restrictive nature of these tools means that generally speaking, users won't be able to break the business applications they're creating. On the negative side, the lack of third-party code will result in a more vanilla application.

Low-Code

Break out your keyboard. As their name implies, LC tools allow their users to add custom code if they like, but with restrictions typically around the following categories:

> The type of code
> The amount of code
> The location of the code
> The capability of the code

As a class, LC is more permissive than NC. The former isn't locked down to the same extent as the latter, so low-coders can customize their applications more. Forget merely changing colors or fonts à la Myspace back in the day; adding entirely new features and integrations with third-party apps is all on the table.

On the flip side, low-coders can easily break the business applications they're creating, especially if they don't know what they're doing. There are too many examples to cite, but here's a note of caution: A single misplaced, extra, or missing bracket, semicolon, or slash in the code can render a custom application or website inoperable. In this way, low-coders resemble proper software developers. Precision matters.

Note that both low-code and no-code are misnomers. Proper developers at all software vendors write thousands of lines of code to make LC/NC happen. Think of it this way: For software to work, *someone* has to write the code. The only question is who. It's a simplification, but Figure 3.2 shows the preferred tools of the three types of developers.

Primary Tools by Type of Coder

| No-Code | Low-Code | Full-Code |

Figure 3.2: Primary Tools by Type of Coder

In other words, if you're using your mouse, it's because software developers used their keyboards.

Shoe Sizes and Standards

Say that you wanted to buy a new type of shoe in your size (women's 8) in the US from the online retailer Zappos—owned, of course, by Amazon. Will your new shoes fit when they arrive?

Short answer: Probably, and for that you can thank the International Organization for Standardization. The body promulgates standards in all sorts of industries and areas, including medical devices, automobiles, and—yes—shoe sizes.[20] (How can you *not* love an organization whose tagline is, "Great things happen when the world agrees"?)

Many emerging technologies and tools, however, lack agreed-upon standards—and low-code/no-code falls into this group. What's more, this void can confuse consumers. For instance, software vendor X may market its main product as low-code. Its competition may sell a similar offering but deliberately opt not to promote it as such.

The Major Characteristics of Today's LC/NC Tools

The start of this chapter covered some popular 1990s visual programming tools. To be sure, many of the business applications created solved real problems at the time. (In some cases, they remain alive and well today.) As a lot, however, nascent LC/NC tools—and their progeny—simply can't compare to today's cohort. At a high level, the latter are an order of magnitude more potent than the former—and there are far more of them. Figure 3.3 displays the comparison between the two eras.

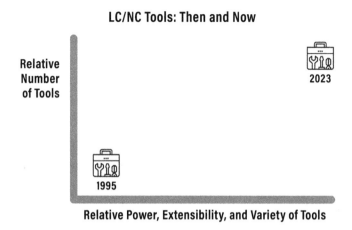

Figure 3.3: LC/NC Tools: Then and Now

A Quick Disclaimer

Before continuing, a disclaimer is in order. All business tech isn't created equal—and no sane person should claim otherwise. This chapter specifically focuses on the business applications that nondevelopers can build with LC/NC tools. I'm intentionally ignoring expensive enterprise-grade business systems and applications—the ones that have historically required massive capital outlays and teams of IT professionals.

I do this based on a sense of equity. There's no sense in trying to compare:

> A single no-coder or low-coder without any programming background; and

> A group of thirty proper software developers, consultants, and internal IT folks.

Apples and coconuts.

Here's a different analogy: No matter how promising, an eight-year-old tennis prodigy won't give Roger Federer a run for his money anytime soon.

Power, Utility, and Breadth

Returning to Figure 3.3 for a moment, the LC/NC tools of the late 1990s don't resemble those of today. As we'll see throughout this book, the current generation of tools lets nondevelopers build an array of über-useful business applications. No, their apps can't split the atom, but make no mistake: They're solving critical problems that plague organizations of all sizes. Moreover, they're not just for web designers and business analysts looking to track simple information in a spreadsheet.

Nontechies can now build and deploy fully functional business apps solely by using LC/NC tools. They can drag and drop. Figure 3.4 demonstrates this tradeoff: The more they use the mouse, the less they have to code.

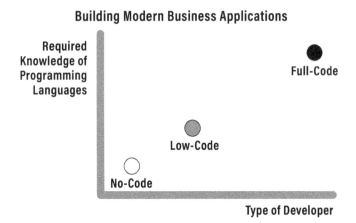

Figure 3.4: Building Modern Business Applications

Despite the power, utility, and breadth of today's LC/NC tools, one truth remains firmly intact: No single instrument can do everything. Even mighty Microsoft can't create an omnipotent business application that meets every employee's need in every company. And that leads me to selling point number two.

Extreme Extensibility

Say that you own and run a small business. Like hundreds of thousands of others, you use Intuit's QuickBooks for essential accounting purposes. (If you're curious, I prefer Wave, but we'll stick with QuickBooks here.)

Like all mainstream accounting packages, QuickBooks' users can quickly enter and pay vendor invoices, run profit-and-loss statements, and much more. That's all well and good, but you wish it could do more than just the bare minimum, maybe even something a little high-tech and exotic. For starters, you despise duplicate data entry for yourself and your staff. Specifically, you want to do the following:

> Use Microsoft Excel to read data from—and write data to—QuickBooks. Examples include adding new customers, vendors, sales, and accounts.
> Easily capture employee out-of-pocket travel expenses, enter the related transactions, pay bills, and reconcile those payments with your business credit card. Ideally, your employees can continue using Expensify.

Now, QuickBooks doesn't *natively* perform these tasks as of this writing. Fortunately, you don't need to start researching new accounting applications or hire an expensive developer to build a custom app. Intuit offers many valuable add-ins and integrations for users to *extend* QuickBooks' out-of-the-box functionality.[21] Anyone remotely proficient with computers can integrate Microsoft Excel,[22] Expensify,[23] and many other useful third-party apps within minutes and for a minimal charge.

In the last paragraph, I deliberately chose the verb *extend*. The main point is that, by default, any business application or system allows its users to perform certain functions. Not a single one, however, meets every conceivable business need. The requirements for a large trucking company differ from those of your local dentist or florist, never mind a large bank like Chase, an auction site like eBay, and so on. To this end, over the years, software vendors have increasingly embraced *extensibility*: the ability to augment native functionality of a piece of software.

Depending on the specific program, the enhancements go by different monikers: *apps, integrations, add-ins, plug-ins,* and *extensions*. The names may differ, but all offer the same fundamental benefit: letting third-party developers and firms extend the core features of a business system or application. As the following sidebar demonstrates, Intuit is hardly alone in understanding the power of developers, communities, and ecosystems.

Extensibility Drives the Remarkable Popularity of WordPress

WordPress is no niche content management system. The world's most popular CMS would never have approached that number if it couldn't meet a diverse set of business needs.

Founding developer Matt Mullenweg embraced the ethos of open-source software from the get-go. Since WordPress's inception in 2003, users and developers have been able to customize it—and many have. As of this writing, users can download or purchase a mind-boggling 55,000 plug-ins[24] and 31,000 themes.[25]

An increasing number of those themes are—you guessed it—of the LC/NC variety. Since 2014, I've used one of them. More than 800,000[26] people like me drag and drop with the Divi WordPress theme from Elegant Themes. (Figure 3.5 shows a screenshot of it.) I'm curious and technical enough to insert different code snippets throughout my site. As an admitted geek, I've added plenty of jQuery, CSS, JavaScript, and PHP to my site over the years. I'm more of an LC than an NC guy.

Nontechies can stay in the NC lane. They can create a visually pleasing, fast, and responsive website without adding a lick of code.

Divi WordPress Theme

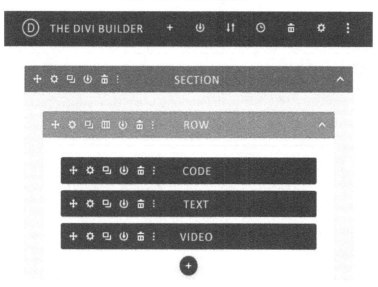

Figure 3.5: Divi WordPress Theme, Source: Elegant Themes

Interoperability on Steroids

Extensibility relates closely to its cousin, interoperability: the notion that Application A can seamlessly talk to Application B. Lest you think that it's a new concept, I can assure you it's not. What's more, as the two examples in the following sidebar illustrate, it's been around for a long time.

Vendor-Based Interoperability

Since the mid-1990s, organizations have found the enterprise resource planning systems mentioned in Chapter 2 useful because of their tight integration. Say that a large pharmaceutical company runs the Oracle E-Business Suite. Employees in different departments access different applications, modules, reports, forms, and information, but all exist in the same underlying ERP system. Exhibit A: Employees in accounts payable cut checks for vendors, and the payroll department did the same thing for employees. Regardless of who did what and when, all of those financial transactions hit the same general ledger. No one wants those checks to bounce.

Outside of the ERP world, the level of integration within programs in the Microsoft Office suite has been remarkable for decades. It's easy to create custom letters in Word and emails in Outlook from an Excel spreadsheet. On a different level, I created dozens of souped-up Microsoft Access databases during the aughts. I would then schedule reports to run in the wee hours of the morning. My clients would turn on their computers, launch Outlook, and find fresh emails with Excel workbook attachments in their inboxes.

Although impressive at the time, these two examples of interoperability are table stakes today. It's now fair to ask: Why *wouldn't* all native Oracle programs in the same suite talk to each other? Ditto for the suites of Microsoft, SAP, Google, and other prominent software vendors.

Now that we've introduced the fundamental concepts of extensibility and interoperability, let's return to the LC/NC world. In terms of depth and breadth, today's tools are far more interoperable than their predecessors. At a high level, they deeply integrate with services, applications, and other tools from a variety of software vendors. In other words, contemporary LC/NC tools tend to play nicely with a swarm of other third-party applications.

For instance, today anyone with zero coding chops can do the following:

> Embed a Google Sheet or YouTube video in Notion.
> Create a channel-specific Slack notification when someone enters data in Airtable.[27] (No sense in cluttering your inbox with dozens of emails.)
> Automatically capture and send tweets containing a specific keyword to Microsoft Power BI.[28]

I could keep going, but you get my point: If you're attempting to accomplish a relatively common task, you can assume that someone else has faced a similar dilemma. With a little digging, you can find a free or affordable integration requiring zero or minimal code.

Even better, if there's no native integration, a powerful LC/NC automation tool from Zapier, Workato, Make, or IFTTT can probably get the job done if the organization is willing to change.[29]

Shareability, Portability, and Collaboration

The apps built with early LC/NC tools by single users worked best for—wait for it—single users. As such, they existed in a relative vacuum: employees creating standalone business applications designed to work on *their* proper computers. The process of supporting multiple concurrent users alternated between clunky and nonexistent. Portals, shared drives, and corporate intranets could only do so much.

In this era, deploying the app over the web wasn't a given; it was usually an afterthought. Collaboration among teams and departments *within* the same organization was maddening, never mind between and among different ones. Data-synchronization issues were rampant. (To be fair, even certain enterprise apps back then required employees to replicate their data when they reconnected to their networks.)

Today, it's a whole new world. Advances in cloud computing, broadband, and mobile devices have untethered users from their proper computers and offices. Not coincidentally, today's LC/NC tools are almost always device- and location-agnostic. They tend to work anywhere on any device. As long as users are online, they can perform largely the same tasks no matter where they are. Finally, and most critically, users rarely have to tell their colleagues to get out of the application—or kick them out—so they can perform necessary updates. Concurrent users are the rule, not the exception. That is, people who use these tools can quickly share their apps with others within their employers' walls—and outside of them. (Chapter 8 discusses the overarching LC/NC philosophies that organizations can adopt.)

Performance, Scalability, and Data Storage

Users can easily add new features and multiple users to apps created with LC/NC tools. That's great, but we must answer equally essential questions:

> Will the extra data that other users create cause these apps to collapse at scale?
> Will the business LC/NC apps eventually burst at the seams, or will they be able to hold up over time?

Twenty years ago, these were valid concerns. Longtime users of Microsoft SharePoint know what I'm talking about.

Generally speaking, today's LC/NC tools scale very well. (Yes, it's necessary to qualify the previous statement. Making blanket statements about them is downright irresponsible.)

Flexibility

As you'll learn throughout this book, today's breed of LC/NC tools can easily serve multiple purposes. Chapter 4 attempts to place them in separate buckets, but their expansive nature makes doing so problematic.

Popularity

"You are not alone."

When comedian Gary Gulman uttered these words, he wasn't talking about today's popular LC/NC tools. In his hysterical 2016 special *It's About Time,* he was referring to his recent experience at Trader Joe's in Manhattan.

In the context of this book, however, Gulman's words ring true. Millions of people use the tchotchkes described in the next chapter

---

* Currently on Amazon Prime Video.

to solve all sorts of problems. Their popularity means that users can access robust communities on Reddit and Discord, Facebook and LinkedIn groups, free and affordable templates, and loads of courses. (These resources are essential, as IT departments generally don't support employee apps built with LC/NC tools. Chapter 6 will revisit this topic.)

Affordability

Chapter 2 covered the software-as-a-service business model. With rare exceptions, the major LC/NC vendors discussed in Chapter 4 make their wares affordable to the masses in this way.

The ramifications of SaaS are enormous: An individual, team, or firm need not make an expensive, multiyear commitment to a tool before knowing if it will stick. Think dating, not marriage; I'm talking a monthly Netflix subscription, not a two-year AT&T or Verizon contract. It's typically easy to upgrade and downgrade plans as needed. Ditto for adding users to existing plans.

As a bonus and to remove any remaining semblance of fear, LC/NC vendors tend to offer relatively generous free plans. Users can play around before becoming proper customers and paying reasonable monthly fees.

Rollbacks and Version Control

The enterprise-grade systems described in Chapter 2 have long shipped with rollback and version-control functionality. It's no understatement to call each of these features *critical*. Apps built with LC/NC tools generally contain identical functionality, although you may need to upgrade to a vendor's specific plan to enable it.

Nondevelopers can effectively click the Undo button after realizing their mistakes, but it gets even better. To the extent that others will be using their apps, restoring them to their prior, error-free states can save enormous time and rework. (Yes, I'm speaking from experience with my own creations.)

I was curious about how frequently Airtable citizen developers used this function, so I created a poll on the r/Airtable subreddit.[30] Nearly 60 percent of respondents reported using backups regularly. By the way, Airtable isn't unique. Smartsheet, Spreadsheet.com, and Notion are just some of the vendors covered in Chapter 4 that offer this valuable feature.

Ability to Rapidly Deploy

Developing a single-person business app is one thing. Deploying it for others to use, however, has historically posed problems. Depending on an organization's level of bureaucracy, it involved coordinating with IT, holding a few meetings, and filling out change requests.

Today's LC/NC tools, however, allow users to rapidly launch their new business applications if the organization allows it. (Chapter 8 discusses LC/NC management philosophies in more depth.) Nondevelopers can release apps for colleagues, partners, and customers to begin using immediately.

A Quick Note for the Low-Code/No-Code Skeptics

Maybe you're still a bit skeptical at this point. I mean, come on. How valuable can LC/NC and their offspring really be?

I've seen people foolishly dismiss this group of tools throughout my career. Don't make that mistake. They frequently serve small but critical business functions—especially when organizations face tight deadlines. The following sidebar illustrates just one simple example of low-code/no-code in action.

Low-Code/No-Code to the Rescue

During the aughts, I made my living helping organizations implement enterprise resource systems. In 2004, South Jersey Gas contracted me for four months to help it go live with the Lawson ERP suite.

The project was proceeding reasonably well as our go-live date loomed. A few days before we planned to activate Lawson, SJG's payroll manager discovered a critical issue with PR132—the overtime calculation program. No one noticed the problem during system testing and training.

I'll spare you all the deets, but the headline is that PR132 calculated employee overtime in a way that complied with New Jersey labor laws. SJG, however, had negotiated terms with different unions promising to pay them *more* in certain situations than the law mandated.

The utility had to pay its employees correctly. Failing to do so would result in a bevy of union grievances—something everyone wanted to avoid. The project management office (PMO) convened to discuss the showstopper and two options on the table.

First, we could push the go-live date, but executives declared that option a nonstarter. Second, we could customize the PR132 program. IT would have to indefinitely test all future system patches and upgrades to ensure that the changes stuck and nothing else broke. Better, but not great.

I suggested a third way. I could put on my citizen developer hat and build a simple Microsoft Access application. It would read the original data, modify it, and spit out a properly formatted export file. The IT manager could then import the revised file back into Lawson and proceed as usual.

The powers that be instantly approved the idea. I then created the little Access app. I tied it to a simple form, tested it, and created a brief guide on how to use and maintain it. The app worked, and we hit our go-live date.

There's an interesting postscript to this story in Chapter 11.

Chapter Summary

> Microsoft Access, FrontPage, and WordPress were some of the first LC/NC offerings.
> Compared to their predecessors, today's LC/NC tools offer unprecedented extensibility and interoperability.
> Affordability, shareability, and easy deployment are other LC/NC benefits. Nondevelopers can roll out these tools quickly and sans IT involvement—if the organization permits it.

>_Chapter 4

The Current LC/NC Market and Vendor Landscape

"All models are wrong, but some are useful."

—GEORGE E. P. BOX

The previous chapter provided the background for contemporary low-code/no-code. It also listed the significant benefits of low-code/no-code, but in general terms.

This chapter serves two purposes. It begins by describing the size and dynamics of the low-code/no-code market. (Hint: It's neither small nor static.) It then covers some of the most popular LC/NC offerings.

A Burgeoning Industry

For starters, the alarming shortage of tech professionals detailed in Chapter 1 is anything but a recent occurrence. The years-long dearth of developers has forced countless organizations to reevaluate their stances toward shadow IT and adopt LC/NC tools. As

69

a result, the LC/NC market has grown considerably over the past decade, but don't take my word for it.

Consider the words of Fabrizio Biscotti, a research vice president at Gartner:

> While low-code application development is not new, a confluence of digital disruptions, hyper-automation, and the rise of composable business has led to an influx of tools and rising demand.[1]

Rising demand is code for what the kids call *mad stacks*. As Figure 4.1 displays, the total LC/NC market should crack $15 billion by the time this book drops.

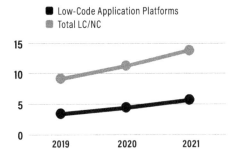

Revenue From Low-Code Development Technologies (Billions of U.S. Dollars)

Figure 4.1: Revenue From Low-Code Development Technologies (Billions of US Dollars), Source: Gartner

Generally speaking, however, organizations tend not to spend billions of dollars on new technologies without—you know—*using them.* Zuck's metaverse is the exception that proves this rule.[2] The massive investment may turn out to be prescient a decade from now, but the current verdict is mixed at best. One tweet calling his avatar "eye-gougingly ugly" received more than 30,000 likes.[3]

Returning to our current, physical world, it's fair to say that LC/NC tools aren't exactly collecting digital dust. By 2025, Gartner predicts that a full 70 percent of new applications developed by enterprises will use low-code or no-code technologies. In 2020, that number was less than 25 percent.[4]

Acquisitions

These are impressive numbers, but there's another, arguably more interesting, way to view the importance of low-code/no-code: Look at what the current tech behemoths are doing.

In 2018 and 2019, AppSheet was generating quite the buzz as a leading LC/NC tool. Forrester Research had given it massive props. On January 14, 2020, Google gobbled it up for an undisclosed sum. As Ron Miller wrote for TechCrunch:

> With AppSheet, Google gets a simple way for companies to build mobile apps without having to write a line of code. It works by pulling data from a spreadsheet, database, or form and using the field or column names as the basis for building an app.[5]

Not to be outdone, SAP acquired the NC app builder AppGyver less than one month later. I'm not privy to the prices that Google and SAP paid for their respective acquisitions, but I'd be shocked if each didn't exceed $50 million—and I'm probably being conservative.

The Innovator's Dilemma

Lest you think that each acquisition represents the acme of tech largesse, I submit that there's a method to each vendor's madness. Despite their mammoth sizes, the two companies—and the rest in their cohort, for that matter—fear disruption. And this leads us to Clayton Christensen's seminal 1997 text, *The Innovator's Dilemma*.

At a high level, Christensen distinguished between two categories of innovations: sustaining and disruptive. In the case of the former, laptops followed desktop computers and expanded the existing market for PCs. As for the latter, the early Netflix DVD-by-mail offering disrupted—and ultimately decimated—Blockbuster and Hollywood Video. It wasn't long before other similar brick-and-mortar outfits tumbled. RIP, Tower Records and Borders Group.

In hindsight, we can determine which innovations fell into each of Christensen's classifications. (They make for great fodder in MBA classes.) Senior leaders, however, lack that luxury. These folks have to make large, real-time bets on which is which. Execs who pooh-poohed nascent ecommerce may have felt vindicated when the dot-com bubble burst. Those who continued to deny the convenience of web-based shopping, however, generally didn't last long.

Let's bring this back to low-code/no-code. Senior leaders at prominent software vendors clearly recognize the import of this potentially disruptive category. Rather than pretend that the world will stay the same, they're betting on change. Better to be in front of it than behind it and risk becoming the punchline to a business joke.

Major LC/NC Subcategories

Given the growth of LC/NC tools, a book of any reasonable length can't list them all. Even if an author somehow magically turned this trick, the list would be outdated before the book hit the shelves.

Still, it's natural to wonder about the general types of LC/NC tools. Which specific ones are nondevelopers currently using to build their bespoke business applications and lightweight systems?

Disclaimers and the Great Classification Dilemma

Before continuing, here are a few notes about my methodology.

First, I visited the vendors' websites to see how they described themselves at the time of this writing. I then placed the LC/NC tools into *primary* categories. Creating an accurate taxonomy proved especially vexing because these tools tend to do several or even many things out of the box—and well.

airSlate epitomizes the widespread tool overlap among categories. (Full disclosure: I've previously worked with the company.) On its website, the company bills itself as "the first and only holistic no-code business automation platform." Even if that were true, what does that mean on a practical level?

In short, quite a few different things. airSlate offers different types of automation around document routing, forms, signature collection, chatbots, and more. It's impossible to place it in a single category.

Formstack—another company with which I've previously worked—also fails to stay in a single lane. As its name implies, Formstack lets users quickly build web-based forms to collect information, but that's hardly all. Its offerings resemble those of airSlate. Think of the following sections as starting points, not as definitive, all-inclusive reductions of these vendors' tools.

I'd be shocked if execs from some of the vendors detailed in this chapter *don't* object to the following taxonomy. In response, I'll steal the timeless quote from the great statistician George E. P. Box at the start of this chapter.

The following vendor list isn't remotely comprehensive. A graphic of just the LC/NC *startups* isn't for the visually challenged.* And good luck keeping it current. By the time you read these words, at least a few of these upstarts won't exist in their current form, if at all. Finally, the forthcoming vendor names and offerings may

---

* See https://tinyurl.com/LC/NCstartups.

seem overwhelming. I get it. Chapter 9 provides guidance on how to evaluate different LC/NC tools.

With all those disclaimers out of the way, let's light this candle.

Flexible Swiss Army Knives

Want to add a little panache to your garden-variety Microsoft Word or Google document? Think of this crop of LC/NC tools as traditional documents on steroids. Anyone can quickly create easy-to-navigate company wikis and rich multimedia documents.

Examples include Notion, Almanac, and Coda. In July 2021, the latter closed $100 million in funding at a $1.4 billion-dollar valuation.[6] Microsoft's one-year-old entry Loop not so subtlety attempts to ape the features of these tools.[7]

Speaking of functionality, I haven't attempted to describe all the native features of these LC/NC tools for the sake of brevity. Now toss in the extensions mentioned in the previous chapter. (See the section "Extreme Extensibility" in Chapter 3.) It becomes impossible to accurately and completely classify Notion, Coda, and the like.

Let me put it this way: Claiming that these tools only let people take notes is akin to saying that a Tesla only holds coffee cups.

Work and Project Management

The same challenge exists with the tools in this adjacent category of work and project management: Smartsheet, ClickUp, Plutio, Quickbase, SmartSuite, Atlassian's Confluence, Wrike, Teamwork, and monday.com. Yes, employees use these applications to manage projects, but each offers far greater customizability than traditional PM tools. This new cohort allows employees in marketing, sales, HR, and other departments to create custom applications that wouldn't necessarily fall under the PM umbrella.

Multiuse App Builders

Glide lets anyone "turn spreadsheets into software."[8] For its part, Draftbit provides a "flexible way to visually create, customize, and launch native mobile apps."[9] Stacker, Buildbase, Zoho, Retool, Adalo, Blaze, Bubble, Appy Pie (great name), Caspio, and Mendix are just a few of the alternative offerings.

If your organization subscribes to Google Workspace or Microsoft Office 365, check out the lesser used arrows in those quivers. For instance, Microsoft Power Apps lets nontechies build bespoke business apps, as we'll see in a Chapter 7 case study.

Automation and Chatbots

Looking to save time and reduce manual work? Zapier, Workato, UiPath, IFTTT, airSlate, and Make (formerly Integromat) are just some of your options. For automation within internal collaboration hubs, consider Microsoft Power Automate (with Teams) and Slack's Workflow Builder. At present, the last two are diamonds in the rough. I suspect that Microsoft and Slack user data would confirm my suspicion: Relatively few people use these powerful LC/NC automation tools.

Automation can make manual, standard business processes more efficient, but what else can it do? What about streamlining a conversation—or at least the part of it that makes sense?

A few years ago, I noodled with Landbot and found it intuitive. After building a decent chatbot, I inserted the code on my website's speaking page. Evidently, plenty of others shared my experience. Two years ago, the company landed $8 million in Series A funding.[10] Landbot is hardly alone. Its competition includes Juji, Intercom, and Drift.

Form Builders

To one extent or another, all the vendors in this category offer ways to automate business processes and workflows. (I avoid using the latter term as much as I can because it reeks of jargon.)

Peccadilloes aside, the impressive tools in this class put Google Forms to shame. The aforementioned Formstack lets people create advanced no-code forms for documents, eSignatures, and more. Sitting next to it are Typeform, Cognito Forms, JotForm, and scores more.

Commerce, Payments, and Transactions

Swan bills itself as "the easiest way to embed banking features into your product."[11]

Great, but is that claim true?

Who knows? I suspect, though, that the marketing bigwigs at Solaris, HUBUC, and Orenda would dispute that contention. It turns out that a slew of fintech startups are working on *embedded finance*. McKinsey defines it simply as "banking-like services offered by nonbanks."[12] Think of it as the lovechild of finance and ecommerce.

We will always need banking, but maybe not banks.[13] If embedded finance ultimately becomes a thing, low-code/no-code will play a significant role.

Data

The sheer variety and flexibility of data-related offerings make them difficult to categorize. These tools allow anyone with respectable tech skills to quickly create lightweight systems and useful apps, some of which overlap with the other categories in this section.

Spreadsheets on Steroids

Spreadsheets on steroids is the best way to describe Airtable, Google Tables, Smartsheet, Rows, Spreadsheet.com, Grist, and their ilk. Anyone with an even cursory understanding of databases can create web-based applications that track gobs of information. End users benefit also. Forms, Gantt charts, Kanban boards, grids, and calendars are just some of the different layouts available. Finally, these tools ship with robust automation features.

If all this functionality sounds incredibly useful, trust your judgment.

Analysis

Plenty of other software lets nondevelopers easily store, analyze, manage, manipulate, and visualize data. Additional examples include Microsoft Power BI, Tableau (now part of Salesforce), Google Data Studio, and Domo's "data apps."[14]

Websites and Content Management Systems

This category is perhaps the most mature one. WordPress, Squarespace, Wix, and Webflow are just a few of today's popular LC/NC website builders. As a group, these vendors make it easy for their users and customers to add integrations and features without ever touching their keyboards.

Contextualizing Low-Code/No-Code

A little context on LC/NC is in order before wrapping up this chapter.

I've just listed dozens of behemoths, emerging vendors, and startups. Each is honing its LC/NC products and services every day. The offerings in the previous section aren't close to comprehensive. Now, remember the high-level stats at the start of this chapter and the premise of this book: LC/NC represents the future of business apps.

Add it all up, and you might think that *every* software vendor has found the LC/NC religion—or will soon. It's only a matter of time before all knowledge workers will *have* to build their own apps, whether they like it or not, right?

Allow me to allay your fears. Yes, LC/NC is quickly rising in popularity, but it won't supplant all the systems described in Chapter 2—and the data supports that assertion.

Let's return to Figure 4.1 for a moment. Gartner pegged global spending on enterprise software alone at $600 billion in 2021—40 times as high as its LC/NC estimate. For additional context, consider that worldwide IT spending in 2021 was roughly $4.25 trillion (USD).[15] LC/NC represents one-third of one percent of this mind-boggling amount.

Don't expect LC/NC spending to surge past IT's expenditures on security, data centers, enterprise software, or other stalwarts anytime soon. Chapter 11 provides several additional rationales for why the clock on traditional business applications isn't exactly ticking.

Chapter Summary

> Low-code/no-code has arrived in earnest. Today an astounding array of user-friendly and affordable tools makes it easier than ever for nontechies to build and launch their own business applications.

> It's not 1998. Whether you want to collect or track information, create an engaging set of documents, manage your workforce, or accept payments, people no longer need to learn how to code.

> The market for LC/NC tools is rapidly maturing. Not every scrappy startup will survive.

Chapter 5

Introducing the Citizen Developer

"You know the business, and I know the chemistry."
—BRYAN CRANSTON AS WALTER WHITE, *BREAKING BAD*, "PILOT"

The low-code/no-code movement is democratizing the process of developing business apps. Throughout this book, we'll see how all sorts of organizations are using them to solve critical business problems.

Skynet hasn't arrived yet, though. These newfangled business applications and lightweight systems don't magically emerge by themselves. With traditional IT departments struggling to meet employee demand, who will build them?

This chapter answers that question in spades. It introduces an emerging and different breed of app maker: the citizen developer.

History and Definition

Until now, I've intentionally described the people who build apps with LC/NC tools as nondevelopers, no-coders, and low-coders.

That is, I've specifically avoided the term in this book's subtitle: *citizen developer*.

I won't take credit for inventing the name. A prominent research outfit coined it more than a decade ago. (In this way and as Chapter 3 describes, citizen developers parallel the origin of *low-code* tools.) I'll cede the floor to Gartner's Ian Finley, author of the 2009 report "New Developers Can Help Deliver More."* In his words:

> A citizen developer is a user who creates new business applications in partnership with corporate IT. In the past, end-user application development was typically limited to single-user or workgroup solutions built with tools such as Microsoft Excel and Access. However, users now can build applications that serve a virtually unlimited number of users without IT's help or knowledge. IT can encourage user developers to come out of the shadows and be citizen developers by providing them with adequate support.

Despite its age, Finley's definition holds up well today. I'll refine it just a skosh and define *citizen developers* as follows:

> People who can build powerful business applications using contemporary LC/NC tools. Their lack of prior programming knowledge and experience doesn't impede them from contributing in the same ways that traditional software developers do.

Citizen developers differ from proper software engineers and coders in several key areas. Table 5.1 summarizes them.

---

* You can view it at https://tinyurl.com/ps-ian-lcnc.

Typical Characteristic	Traditional Software Developer	Citizen Developer
Background	Software engineering or computer science	Varied
Department in organization	Information technology, information systems, or DevOps	A functional area, such as marketing, sales, HR, or finance
Functional knowledge of specific business domains	Low	High; these peeps are usually subject matter experts

Table 5.1: Major Differences Between Traditional Software Developers and Citizen Developers

Gartner's current IT glossary makes a valuable point about contemporary citizen developers. Specifically, the "citizen developer is a persona, not a title or targeted role. They report to a business unit or function other than IT."[1]

It's important to unpack that particular nugget. *Citizen developer* isn't a terribly common job title, at least for the time being. A quick search on indeed.com confirms as much. (One at the Kennedy Space Center in Florida* and another at Comerica Bank in Michigan† serve as exceptions to that rule.) Instead, think of citizen developer as an increasingly desirable *attribute* of an existing position. When describing citizen developers, a few job postings include language to the effect of "otherwise known as business technologists." Interesting point.

You'll meet some actual citizen developers later in this chapter. For now, here's a simple persona for a modern one:

Elaine is a twenty-seven-year-old senior marketing analyst for a retail company in Portland, Oregon. She studied business and

---

* See https://tinyurl.com/lead-citdev.
† See https://tinyurl.com/comerica-cd.

communications in college. Elaine uses several LC/NC tools
in her job to collect data, automate manual processes, and
communicate with her colleagues. Her more senior colleagues
frequently come to her with technology questions, and she's
happy to help them.

Let's move on to an interesting query: If citizen developers
harken back to 2009, why are they exploding now?

The Rise of the Citizen Developer

Returning to Chapter 1 for a moment, the labor market for tech
talent has been scorching for years with no end in sight. The pandemic and the subsequent move to widespread remote and hybrid
work exacerbated the issue. Chief information officers (CIOs), IT
directors, heads of other functions, small business owners, and
others face an insatiable demand for business applications.

Against this backdrop, these people have begun raising the
white flag. They're surrendering the war on shadow IT they've been
fighting for years. In the process, they're increasingly welcoming
technologies that employees from different lines of business build,
buy, rent, and use.

Consider some fascinating statistics and predictions from
Gartner:

> By 2023, the number of active citizen developers at large
enterprises will be at least four times the number of professional developers.[2]

> And this group will be busy. Citizen developers will build
the majority of technology products and services by 2024.[3]
(Adding fuel to the fire, Microsoft predicted in 2020 that

citizen developers would build 450 million apps by 2025 using LC/NC tools.[4])

> As of this writing, business-led IT practices represent roughly 36 percent of IT budgets—a number that has steadily increased in the past few years. The trend should continue.

> By 2023, citizen developers will build four in five business apps.[5] Figure 5.1 shows this breakdown.

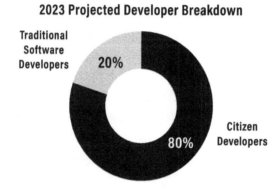

2023 Projected Developer Breakdown

Figure 5.1: 2023 Projected Developer Breakdown, Source: Gartner

In June 2021, Forrester Research released the report *Low-Code Platforms for Business Developers, Q4 2021*. The company conducted an extensive survey of proper senior-level techies. More than a quarter of respondents reported that "their organization develops applications using citizen developers and low-code platforms."[6]

The move toward the widespread internal acceptance of LC/NC tools is unmistakable. Case in point: Gartner predicted at its annual IT Symposium in October 2020 that by the end of that year, at least 70 percent of large enterprises would have formalized policies around citizen development and vendor management. The corresponding number a decade ago was a mere 20 percent.[7] Five dollars gets you ten if that number doesn't jump to 80 percent by 2025.

Opportunity in Chaos

In a way, Figure 5.1 understates the speed and results of the flywheel. Let's not forget the bevy of small businesses—nearly 32 million alone in the US as of this writing.[8] More than one in four small businesses lacks formal IT support.[9] Perhaps this explains why only 17 percent are satisfied with its current use of technology.[10] I'd be shocked if those two groups didn't significantly overlap.

At least there's an upside to the lack of proper IT support: It frees citizen developers to start building apps. Whether they realize it or not, countless employees in HR, marketing, sales, finance, operations, and the like are citizen developers right now. When they build business apps, no one stands in their way; no CIO or IT director stops them from developing what they need to do their jobs. They can access LC/NC tools without the imprimatur of IT. The affordable nature of these tools—discussed in Chapter 3—only intensifies their adoption and emboldens citizen developers to do more.

Finally, consider independent contractors and freelancers. The US Census Bureau reported that business formation in 2020 grew 24 percent compared to 2019.[11] When you work for yourself, you can noodle with new technologies and land on improved applications, systems, and devices. New solopreneurs can adopt the tech that meets their needs; they don't need to retire legacy systems and change others' antiquated habits.

Tools Mature and IT Comes Around

It's no surprise that the gradual acceptance—and now explosion—of citizen developers has coincided with the development of powerful LC/NC tools, but don't take my word for it. As tech journalist Clint Boulton wrote in August 2021 for cio.com:

... the rise of low-code and other automation tools that help citizen developers build applications is democratizing software development beyond IT's control. Second, supporting this cohort of business technologists rather than cracking down on them can help lighten IT's workload, which has grown unwieldy with the acceleration of digital business imperatives.[12]

IT departments today aren't just increasingly *tolerating* citizen development. *Embracing* is a better word. The result is the virtuous cycle shown in Figure 5.2.

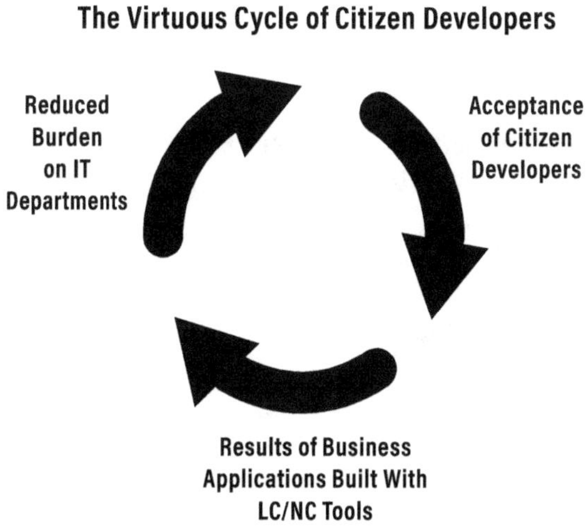

The Virtuous Cycle of Citizen Developers

Reduced
Burden
on IT
Departments

Acceptance
of Citizen
Developers

Results of Business
Applications Built With
LC/NC Tools

Figure 5.2: The Virtuous Cycle of Citizen Developers

Growth begets further growth. Ask ~~Facebook~~ Meta, Amazon, Uber, and Netflix execs how that usually turns out. Network effects in action, baby.

The Pandemic Accelerates the Trend

Finally, consider the pandemic and several of its second-order effects: the proliferation of remote and hybrid work and the Great Resignation. Collectively, they expedited the deployment of LC/NC tools and, by extension, citizen development. In October 2022, KMPG surveyed hundreds of executives across Global 2000 firms in two phases:[13]

> **Phase I:** March–April 2020, 300 executives
> **Phase II**: May–June 2020, 600 executives

Figure 5.3 shows their responses to the question: What is your most important automation investment?

Investment Skyrockets in LC/NC, While BPMS Drops Significantly

Answer to the question: What is your most important automation investment?

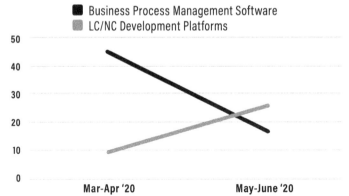

Figure 5.3: Investment Skyrockets in LC/NC While BPMS Drops Significantly, Source: KMPG International

During the early months of the pandemic, investment in business process management software tumbled. Execs reallocated those funds to—you guessed it—LC/NC tools. The data shows that, as a lot, they swiftly realized that empowered citizen developers could at least partially fill the sudden void caused by their employees quitting en masse.[14]

Common Attributes of Citizen Developers

We now know more about the environment that has given rise to citizen developers, but not as much about the group itself. As Jerry Seinfeld says as the punchline to several jokes, "Who *are* these people?"

Background and Technical Chops

In its inaugural 2015 State of Citizen Development Report,[15] Intuit QuickBase surveyed 148 professionals who claimed to be either citizen developers or IT professionals who supported them. Here are some of its findings:

> Nearly all respondents (97 percent) possessed basic word processing and spreadsheet skills. (I'm astonished that that number isn't higher.)
> Thirty-six percent claimed front-end web-development skills in HTML, CSS, and Java. (I'll return to this subject in a moment.)
> A mere 8 percent possessed hard-core, traditional coding skills in Java, .NET, Python, Ruby, PHP, C++, and other modern programming languages.

More recently, IDC estimated in March 2021 that 40 percent of low-coders are also professional developers.[16] Put differently, 60 percent of these people are *not* software engineers—and the reason stems from how the tools have developed in the past decade. Specifically, LC/NC tools have progressed a great deal since 2015.

As the following sidebar illustrates, even the most ambitious low-coders don't have to be proficient in object-oriented programming languages.

Gateway Drugs: How I Became a Citizen Developer

—John Elder, Director of Operations, The Business Blocks

Since I was a kid, business has fascinated me. I began investing at school and started a textbook importing business while at university to get back at the administration for overcharging students. Upon graduating in 2012, I moved into a procurement role and started building my business skills: analysis, negotiation, networking, and problem-solving. I knew I'd start a bigger entrepreneurial venture at some point. Fast-forward a few years, and the opportunity presented itself.

I started The Business Blocks in 2021 to help companies take the pain out of their manual processes. I saw them everywhere. It didn't take a rocket surgeon to realize that bringing new employees, customers, and vendors on board was inefficient. The opportunity for errors lurked everywhere. Because I'm not a programmer, I started exploring how low-code tools could enable automation.

Airtable, Notion, and Microsoft Power Automate were my gateway drugs into the LC world. After I'd obtained enough skills to be dangerous with them, my confidence exploded. At that point, I started considering myself a proper citizen developer. I built many mini applications, including an Airbnb management app, a low-code portal to teach professionals about LC/NC tools, and several that automated manual business processes.

Yeah, I was hooked.

Curious knowledge workers will eventually hit their limits in their jobs. They view having to repeatedly perform manual tasks as inefficient and a waste of their valuable time. They're right. An appalling lack of support and solutions from management and IT creates a perfect storm. To increase their productivity and improve their lives, they'll start exploring. Once they do, they'll soon find colleagues getting results via LC/NC tools. Within no time, they'll go down the rabbit hole.

I should know. That's what happened to me.

If you're an aspiring citizen developer, here's my advice:

- Test, try, and break things. What's the worst that can happen? Delete, repeat.
- Be curious, spend a little money, and start playing with new tools.
- Network and talk to people about their real problems.

Citizen developers looking to build more sophisticated apps should possess at least a cursory understanding of enterprise technology. (They could do worse than reading Chapter 2 of this book for an overview.) At the very least, this knowledge will enhance their credibility should they interact with internal IT employees. What's more, they won't waste their time trying to create complex business applications that even today's most potent LC/NC tools simply can't accommodate—and won't anytime soon. No sense trying to win a drag race driving a Prius.*

---

* Yes, this is a reference to the iconic scene in *Horrible Bosses*.

Function

I was curious about citizen developers' functional backgrounds, but I couldn't locate any recent surveys on the subject. To sate my curiosity, I took to Reddit. In August 2022, I launched a poll on the r/nocode subreddit. Over seven days, eighty-five people responded.[17] Figure 5.4 displays the results of my admittedly unscientific poll.

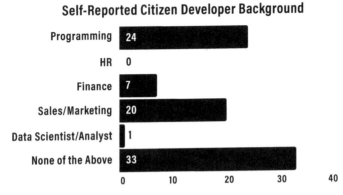

Figure 5.4: Self-Reported Citizen Developer Background, Source: Author, Reddit (N=85)

Looking at the data, immediately two things jumped out. First, 28 percent of respondents reported being proper developers. Along with the far more credible and reliable IDC survey mentioned earlier, one thing is clear: Proper programmers liberally use today's LC/NC tools. Just because software engineers *can* use their keyboard to build apps from scratch doesn't mean they *should*. Sometimes the mouse gets the job done.

Second, 72 percent of respondents identified as subject matter experts, not traditional programmers. More than three in four citizen developers are hybrids; they fuse their functional knowledge with their technical aptitudes to create custom apps.

Size of Organization

Lest you think that citizen developers operate only in conglomerates, think again. In December 2021, researchers from Metropolitan State University found that "citizen developers are real and prevalent across all sizes of organizations."[18]

Age

In March 2014, the software vendor TrackVia surveyed over 1,000 US workers aged eighteen to fifty-five.[19] Interestingly, the company asked self-reported citizen developers if they'd created their own business apps. One could report being a citizen developer but simultaneously *not* be an app builder. (The peculiar distinction hasn't held up well. Mainstream thinking, the majority of contemporary research, and this book consider *all* citizen developers app makers. You can't be one without the other.) Nevertheless, Figure 5.5 shows one fascinating finding from TrackVia's survey.

Age Breakdown of Citizen Developers Who Build Custom Business Applications

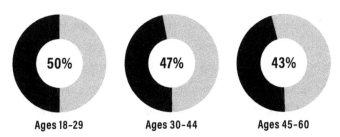

Figure 5.5: Age Breakdown of Citizen Developers Who Build Custom Business Applications, Source: TrackVia Citizen Developer Survey, 2014

Younger respondents were more likely to report building their own apps than their older counterparts.

As I researched and wrote this book, I suspected that citizen developers tended to skew younger. It appears that I wasn't wrong. As for why, here's my working five-pronged theory.

First, Gen Z and millennials are incredibly tech-savvy. They generally embrace new tools and want their colleagues, managers, and employers to do the same. The four years I spent as a full-time college professor confirmed as much. As Gene Marks writes for the *Guardian*:

> Gen Zers bring with them not only a lifelong experience with the cloud, social media, and mobile technology but—because they've known and adapted to this stuff since birth—are also open to more technology, more changes, more adaptations. They expect their employers to embrace the latest if it will help them be productive, independent, and better balance their work and personal lives.[20]

Second, younger workers realize that learning new skills isn't optional; it's necessary to remain employable and pay off crippling loads of student debt.[21] Again, the data reflects as much. A Gallup survey from September 2021 reported that, for young adults, upskilling is the third most important benefit. Only health and disability insurance ranked higher.[22] It even topped paid vacation.

Third, when many corporate employees reach their late forties and early fifties, they're no longer individual contributors; they've graduated to managers, directors, and VPs. As such, they're not as close to—or as hands-on with—the tech as their direct reports.

Next, and as discussed in Chapter 3, even the tech-savviest Gen Xers had to make do with a comparatively limited set of LC/NC tools. (Think Microsoft Access, not Airtable.) Many young citizen developers have *always* used technologies to build and deploy powerful business apps. A large percentage of Gen Xers simply can't make that claim.

Finally, the way that we look at workplace tech is at least partly a function of age. A 2022 survey of 800 employees and IT managers at large organizations reveals the following generation gap:

> Forty percent of millennials and 63 percent of Gen Zers report that their workplace tools are buggy and unreliable, difficult to navigate, or don't integrate well with other technologies. Only 23 percent of boomers share the same concerns.[23]

Senior employees are less likely to report dissatisfaction with workplace tech than their more junior counterparts. It stands to reason that the latter are more likely to become citizen developers. Younger folks are more likely to see opportunities for improvement.

Add all this up, and then consider the following two employees. Which one of them is more likely to be a citizen developer?

> **Jerry:** A fifty-four-year-old VP of strategy.
> **Kramer:** A twenty-nine-year-old senior marketing analyst.

I may be wrong. Equipped with no other information, however, my money is on Kramer. The larger point is that younger workers are generally more curious about technology—and tech-savvy, for that matter—than their older counterparts.

Lest anyone call me an ageist, let me stress this point again: Age doesn't prevent someone from becoming a citizen developer. I'm *not* claiming that more senior people can't become citizen developers. (To say so would be hypocritical; I've considered myself one for decades, and I'm no spring chicken.)

All things being equal, though, it's easier for younger people who lack technical backgrounds to learn new LC/NC tools and become citizen developers. (Chapter 9 covers how to learn new ones. For now, suffice it to say that it's not that hard.)

The main point here is that mindset supersedes date of birth.

Mindset

If there's an objective means to measure curiosity among citizen developers, I'm unaware of it. As someone who has sat at the nexus of business and technology for his entire career, I can attest to the importance of the characteristic. The citizen developers you'll meet in this book will, too.

Kim Bozzella is the head of global technology consulting at Protiviti. Writing for *Forbes* in August 2022, she describes effective citizen developers as those who:

> Need variety in their jobs; they can't work on just one thing.
> Disdain inefficiencies.
> Seek better ways of working.
> Demonstrate a self-driven technical aptitude.
> Exhibit proficiency with LC/NC tools.[24]

Recruiters and hiring managers looking to add citizen developers would do well to screen for these attributes. To this end, an effective behavior-based question is, "Tell me about a time that you used technology to improve an existing business process." Candidates' responses—or lack thereof—will tell you if they walk the talk or are just poseurs.

But let's take a step back. Why should an organization hire a citizen developer in the first place? What does a citizen developer bring to the table?

Funny you should ask. The next chapter answers that question in spades.

Chapter Summary

> Low-code/no-code tools don't build new business applications by themselves. Citizen developers make the magic happen.

> Citizen developers tend to skew younger, but anyone can become one. There's no age floor or ceiling.

> Although extensive technical knowledge isn't a prerequisite, a little nous goes a long way. If nothing else, it helps citizen developers understand the pillars of app development and interact with proper techies.

Chapter 6

The Benefits of Citizen Development

"The difficulty lies not so much in developing new ideas as in escaping the old ones."
—JOHN MAYNARD KEYNES

The Macmillan English Dictionary defines a step change as "a significant change in something, especially one that leads to a noticeable improvement."

Examples of recent step changes in the business world aren't hard to find. Thank technology. Netflix in 1998 introduced a process of ordering DVDs by mail that proved superior to the frustrating brick-and-mortar Blockbuster experience. Ditto for streaming years later. Google's search engine was far faster and more accurate than Ask Jeeves, AlltheWeb, Yahoo!, and the prominent ones of the late 1990s. Uber made ride-hailing far simpler than barking and motioning at taxi drivers from street corners.

No search engine, utility, or streaming service is perfect; all companies and technologies suffer from downsides. In each case, however, we look back and marvel at how much easier their products and services have made our lives. Make no mistake: The same massive leap forward is taking place with business applications.

This chapter fuses the previous two. Specifically, the convergence of low-code/no-code and citizen developers is driving this step change. The following pages cover the significant advantages that organizations can expect to realize by embracing citizen development. Let's begin with the advantages for IT departments and the employees who work in them.

IT Benefits

After years of battling shadow IT, CIOs are increasingly throwing in the towel. They know that, by embracing citizen development, they stand to benefit in the following ways.

Software Developers Can Focus on More Complex Apps

Consider the following scenario.

A team or department needs a new but somewhat limited custom business application or system. Let's assume that the team lead or department head wants to proceed with building. (Remember from Chapter 2 that other viable procurement options exist for new business applications.)

The natural question becomes, Who's going to do the building?

As Figure 6.1 displays, unless outsourcing is on the table, the question is either a proper or a citizen developer.

Let me answer that question with another one: Why tie up a busy programmer on a project that a solid citizen developer can quickly complete using an LC/NC tool?

If pressed, I can posit a few one-off scenarios. Generally speaking, though, the argument against citizen development is a difficult one to make today. The pros far outweigh the cons.

Figure 6.1: Who Will Build the App?

No matter how impressive their chops are, however, citizen developers simply can't build complex, enterprise-wide systems. If someone claims otherwise, run. The greater the complexity of the required business application, the less likely that a citizen developer can build it. Figure 6.2 shows this relationship.

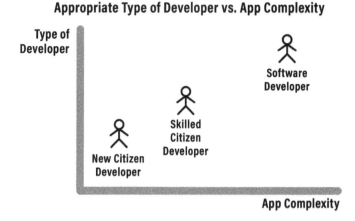

Figure 6.2: Appropriate Type of Developer vs. App Complexity

Citizen developers save IT departments and proper software developers valuable time. Consider the words of Jason English, an analyst and chief marketing officer and analyst at Intellyx. He correctly notes that they can also "lower the barrier to entry for short-staffed IT teams."[1]

Fewer Support and Enhancement Requests

Meet Steve. He works as an accountant and has opened tickets with IT after he's found bugs in his firm's internal systems. Steve also fancies himself a citizen developer and has built his own app. He's unlikely to call IT, however, if his own creation experiences issues. His primary and perhaps only line of defense is the software vendor, not his already overburdened internal support department.

At least Steve isn't alone. Gartner's Jason Wong describes how current IT departments simply lack the bandwidth to learn the ins and outs of approved LC/NC tools, much less the dozens—or even hundreds—of apps that citizen developers build with them.[2] For this reason, the latter will have to learn a few things about application planning, communication, and support. (Chapter 10 explores this subject in more depth.)

Shrink the IT/Business Divide

Getting people within the same department to communicate and collaborate well can be challenging. (See "Understanding the Dual Nature of Contemporary IT" in Chapter 1.) But getting different departments within the same organization on the same page is often an exercise in futility.

In theory, IT works well with its constituents: marketing, HR, sales, finance, and the like. In practice, however, the relationship between IT and different functional areas is commonly strained. Industry types like me call this scenario the *IT/business divide.*

It's a pervasive problem that has plagued companies for decades—even ones that routinely conduct tech-related research and advise clients on how to run their own IT shops.

As Marc Ferranti writes about the hallowed firm Gartner on cio.com in November 2001:

> During the late 1990s, the Stamford, Conn.-based consultancy suffered from a chronic lack of communication between its IT department and its far-flung business units. That disconnect cost the company millions of dollars every year, in the estimation of its own CIO and other observers.[3]

Ironic, right?

My objective in broaching this two-decade-old story isn't to dunk on Gartner. Doing so would be counterproductive. After all, I've cited its excellent research quite a bit in the previous chapters. Rather, this quote demonstrates that the IT/business divide is a tough nut to crack. If it plagued Gartner at one point, why would your organization be immune to it?

I've seen this movie dozens of times in my career. In case you haven't, here's a simple example of the divide in action.

Sarah is a tech-savvy business analyst whose team needs IT to tweak a custom business application. She fills out the formal change request that her IT department requires. A few days later, someone reviews it. Equipped with no other information, I'd bet on one or more of the following events happening:

> The first change won't completely meet Sarah's needs.
> A fair degree of back-and-forth ensues.
> Sarah may not have fully considered other changes in the app that IT will now need to make.
> Those subsequent changes may break other app features.

> The final change(s) will ultimately take much longer than Sarah and her teammates expected.
> She and her team will become frustrated.

Now let's tweak this scenario. What if Sarah could make those changes herself? IT wouldn't have to interpret and prioritize her request in the context of the ever-increasing backlogs discussed in Chapter 1.

Thanks to low-code/no-code, citizen developers like Sarah can fix the application themselves. As a result, the IT/business divide ceases to exist, at least in this specific case. More important, Sarah's team can more quickly access new features of existing business applications and, for that matter, new apps altogether.

Avoid IT Altogether

Equipped with mad Python or Java skills, rockstar developers can automate just about anything.

You may not be one of these people. I'm sure not.

But here's the rub: Citizen developers don't need to be. Their lack of programming prowess doesn't inhibit them from creating apps that meet their constituents' needs. (I'd bet that most end users couldn't distinguish between the apps that IT and citizen developers create.)

Pick a category of LC/NC tools from Chapter 4—say, automation. Zapier, Make, IFTTT, and the other tools in that bucket are easy to learn, tweak, and activate. They generally require minimal ramp-up time, at least at basic and moderate levels. Even better, IT need not develop the app and deploy it.

if they needed to open a case with her. We even recorded a quick video to show people how to use it and posted it in the channel.

Our mini-app wasn't the most sophisticated in the history of Slack, but it got the job done.

And the best part?

No code required. Not a single line.

Nora thanked me several times over the ensuing months for simplifying her life. By the way, you can do the same thing with Teams via Microsoft Power Automate.[7]

Consolidate Existing Workplace Apps and Eliminate Others

Departments, teams, and firms frequently use several disparate, disconnected applications and systems. A 2021 report found that 41 percent of employees reported feeling overwhelmed by the number of tools and technologies they're required to use.[8] I've seen these groups use all of the following tools *on the same project*:

> A Google Sheet to track time
> A proper project management tool
> A separate issue-tracking application
> Microsoft Word to create rich documents
> Email for team-based communication
> A cloud-based storage service such as Dropbox, Box, or OneDrive

Thanks to this new breed of powerful LC/NC tools, however, teams can perform all these functions *within a single app*. ClickUp and monday.com are just two examples of tools that can effectively serve veritable Swiss Army Knives in these cases. Even better, they seamlessly integrate with other third-party extensions. Want to join your 2 p.m. Zoom meeting through monday.com? Knock yourself out.[9]

Beyond consolidation, consider outright elimination. Take Microsoft Excel and Google Sheets, for instance. With some effort, you can hack together passable views of Kanban boards and timelines in these spreadsheet programs. It's not all that hard. Maybe the process involves a browser extension or another third-party app, plug-in, add-in, or service.

By contrast, creating a timeline in Notion is ridiculously easy and requires no code or additional software.[10] Want to create a Kanban board in ClickUp? Go nuts.[11] Doing so is an order of magnitude easier than manually creating one in Excel,[12] and content on its mobile app generally looks great.

Shore Up Data Quality

Poor data quality costs organizations an average of $13 million per year.[13] It leads to questionable business decisions and underlies a firm's attempts to build a culture of analytics.[14]

LC/NC tools ship with robust data-validation functionality that can minimize this scourge. A little data validation goes a long way.

Significantly Improve Existing Business Processes

Speaking of automation, LC/NC tools can significantly reduce the time needed to complete routine business tasks. Researching this book, I discovered the Airtable *BuiltOnAir* video podcast. Figure 6.3 shows a visual from one episode that stayed with me throughout The writing process.

Slack Channel Message Reflecting Process Improvement via Airtable

Two years ago today, I took a workflow from one month to five days. Last year, I cut that five days down to five hours. This year, it takes me 15 minutes.

Figure 6.3: Slack Channel Message Reflecting Process Improvement via Airtable[15]

To quote Larry David, improvements like that are pretty, pretty, pretty good. But what about a specific, concrete example of low-code/no-code tightening things up?

Although I've anonymized the names in the following sidebar, trust me: The situation wasn't hypothetical. Two years ago, Zapier sported more than 125,000 paying customers—a number that has doubtless grown since then.[16]

Zapping Payment Issues

A small law firm has smartly chosen Slack as its internal communication and collaboration hub. The office manager Francesca uses the corporate PayPal account to purchase office supplies and other goods. When a purchase failed because of insufficient funds, she received an email.

Pretty standard, right?

In November, Francesca went to Omaha for two weeks to visit her family. She didn't check her work email during that time. When she returned to the office, essential supplies hadn't arrived because of a slew of payment issues. Her boss Saul was displeased.

To fix the issue, Francesca turned to a little Zapier integration.[17] With a few clicks of the mouse, all PayPal notifications now magically appear in a private Slack channel for Saul to view when she's on vacation and vice versa.

What's more, we can automate work in a way that doesn't erode the human element. In August Christine McHone, Slack Capability Leader at Slalom, appeared on my podcast.[18] She described how Slack's Workflow Builder allows anyone to automate manual tasks in a way that doesn't feel robotic. In fact, the interactions are downright fun.

Expedite Testing New Ideas and Failing Faster

We take the success of Zappos, YouTube, Starbucks, Slack, Netflix, and other multi-billion-dollar businesses for granted. Make no mistake, however: In each case, massive paydays were hardly guaranteed. (For more on this subject, read Duncan Watts's excellent book *Everything Is Obvious: Once You Know the Answer.*)

The mantra in Silicon Valley over the past two decades has been this: If you're going to fail, then fail fast. To this end, why spend a ton of money on a new idea, product, or company if it won't take off anyway? The expense, time, and effort for launching a new tech product service are usually considerable.

Today's LC/NC tools do far more than allow entrepreneurs to create simple websites with contact forms. On the contrary, contemporary low-code/no-code makes launching full-fledged digital products and services possible.[19] The next chapter contains a proper example of how one ambitious and curious individual launched a successful app with Bubble.

For the entrepreneurial reader, just remember this for now: If you're going to fail, you might as well fail fast and inexpensively. Low-code/no-code can be invaluable in this regard.

> ### Elon's Early Coding Adventures
>
> Long before Tesla and SpaceX, Elon Musk cofounded Zip2 in 1995 with his brother Kimbal. Elon handled the coding but was unable to use drag-and-drop tools to create the software for the online city guide. They simply didn't exist, but what if they did?
>
> I suspect that the brilliant and mercurial egomaniac would have considered LC/NC beneath him and banned its use throughout the company.
>
> Too harsh? Read Ashlee Vance's eponymous 2015 book *Elon Musk: Tesla, SpaceX, and the Quest for a Fantastic Future* and decide for yourself.

Individual and Team Benefits

Citizen development does far more than alleviate the burden of overwhelmed IT departments and personnel. We're just getting started.

Quickly Solve Key Business Problems

At a high level, all systems and applications solve problems. Regardless of who built a particular one, when, and with which tools, that general statement holds true.

The same principle applies to LC/NC progeny. They solve longstanding problems and newer ones, as the following sidebar shows.

Automating the Return to the Workplace

This sidebar is a composite based on what several actual citizen developers have built.

Returning to the office has proven to be a logistical nightmare for almost every company, and Omnicorp is no exception. Most of its employees just grin and bear it. After all, aren't the ability to work from anywhere and avoid daily commutes worth a little extra friction? You can't have your cake and eat it too, right?

Marty doesn't think like most people.

He works as an HR analyst at the company and fancies himself a citizen developer. Marty sees the emails that bounce back and forth between HR and facilities personnel. He wonders if there's a better way to handle bookings and reservations. To that end, he starts noodling with ServiceNow, a popular LC/NC solution. Its Safe Workplace Suite seems like just the ticket.

Within two weeks and without tying up IT resources, he creates a skeleton application that lets all Omnicorp employees quickly reserve rooms, equipment, and catering.

But wait. There's more.

Any employee can solicit resources for events, conferences, training, and customer visits from any device. After entering their requests, employees can view and manage them; they don't have to call or email HR or facilities to make changes or inquire about status updates.

Through ServiceNow, Marty has effectively created a bespoke, interactive booking and reservation application. He invites a few colleagues to test it. They identify a few bugs, and Marty quickly squashes them. A month after starting, he

> deploys the app and adds a little custom code. The deluge of employee emails plummets, and Marty receives a promotion.

Reduce Confusion and Miscommunication

The verdict on remote and hybrid work is nearly unanimous: We love it, but there are some downsides. As I wrote in *Project Management in the Hybrid Workplace:*

> The explosion in remote and hybrid work has increased the sheer number of asynchronous, written messages that employees send and receive on any given day. In March 2021, Microsoft reported that the average Teams user sent "45 percent more chats per week and 42 percent more chats per person after hours, with chats per week still on the rise."[20]

Yeah, but we still get to work from home. Don't the fleas come with the dog?

We shouldn't accept as gospel that we will receive a constant stream of messages throughout the day. LC/NC tools let citizen developers build apps that reduce confusion, miscommunication, and the interminable back-and-forth that burns employees out.

Serve as a Valuable Teaching Aid

Current and budding citizen developers don't necessarily possess programming backgrounds. LC/NC tools can serve as a valuable starting point for those who want to take the next step in their careers and build their own business applications. I should know. As the following sidebar illustrates, I began noodling with Microsoft Access in 1998 specifically for that purpose.

My LC/NC Introduction to Databases

In 1998, I accepted a job as an HR specialist at Merck, a large pharmaceutical company headquartered in New Jersey. I wanted to work closer to where I grew up, but another reason prompted the move. I had learned in my prior position that I just wasn't wired for a traditional HR role. As one colleague playfully told me, "We need to work on your warm-and-fuzzies."

She wasn't wrong.

The role at Merck was different; it consisted of a few conventional HR responsibilities, such as recruiting and employee compensation. Sixty percent of my work would be more technical. It would involve HR systems and data analysis. Yeah, that mix was more my speed.

I quickly discovered that my new employer's internal systems and data were, to put it kindly, suboptimal. (If I weren't being so kind, I'd say messy AF.) To effectively do my job, I would have to up my tech game. Although I was proficient in Microsoft Excel, I wanted to learn how to use its more robust suitemate.

Enter Microsoft Access. I bought and read the 1,100-page *Microsoft Access Bible* and followed its detailed tutorials. I enjoyed learning about primary keys, data types, schemas, table joins, Cartesian products, entity relationship diagrams, Structured Query Language, and the other rudiments of modern relational databases. Lightbulbs appeared above my head. More than a few times, I said to myself, "Oh, so *that's* why this happens." I then thought about my current job. I saw inefficiencies and potential solutions everywhere.

I started putting my newfound knowledge and skills into practice. I created simple, standalone databases that helped me identify duplicate or missing records from master lists and Merck's bevy of HR systems.* In hindsight, I'm glad that my early Access databases didn't interact with critical internal systems. A few times at the start of my Access journey, I made mistakes that could have caused all sorts of problems if they had been integrated with larger systems. The training wheels worked.

Although I left Merck in 2000, I didn't leave my brain at the door. The knowledge I gained helped me immensely when I started as a systems implementation consultant. I frequently needed to build mini applications that solved vexing data issues. Many of my database creations allowed nontechnical peeps to run and distribute custom reports for themselves.

I eventually graduated from Access to more robust and sophisticated tools. The applications may have changed, but the concepts I learned a quarter-century ago haven't. They've stood me a good stead.

Let Employees Focus and Compartmentalize

I won't claim to have reviewed every possible LC/NC tool listed in Chapter 4. There are only so many hours in a day. Still, I've noodled with a bunch of them over the years. While researching this book, I experimented with even more of them. Not surprisingly, their features tend to overlap, especially within a specific low-code/

---

* The outer join is a beautiful thing.

no-code category. Exhibit A: Many LC/NC software vendors have embraced the term *workspace.*

A workspace underpins Airtable, Airtable Coda, Notion, ClickUp, Google Tables, and plenty of other LC/NC tools. (In the collaboration universe, Slack and Google Chat have adopted it as well.) Figure 6.4 shows the current ClickUp structure.

ClickUp Hierarchy

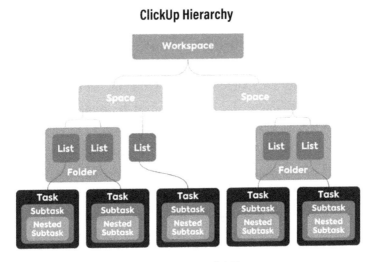

Figure 6.4: ClickUp Hierarchy, Source: ClickUp

At a high level, workspaces allow users to create dedicated containers for each of their projects, clients, teams, departments, and more. With a few clicks of the mouse or taps of their fingers, users can assign different permissions to others, regardless of where they work. Just as important, users can easily customize team-, client-, and project-based notifications in different workspaces.

The idea behind dedicated workspaces is profound. All documents, notes, files, and communications exist in a dedicated container. While working for one client or project, you're far less likely to get distracted by another. (Of course, you may decide to lump everything together into a single workspace.)

Through this type of intentional work segmentation, LC/NC tools allow users to perform deep work—to borrow the phrase that Cal Newport coined in his bestselling book. To ignore a team, client, or project, simply log out of its associated workspace and enable do-not-disturb on your computer. If you don't strictly enforce boundaries when off the clock, you have two options:

> Don't install the LC/NC app on your mobile devices.
> Install the app on your mobile devices, but only log in to your personal workspaces. That is, don't take your work with you.

As an added benefit, these deliberate choices can reduce work-related stress—an alarming problem that remote and hybrid work has exacerbated.[21] McKinsey reported in April 2021 that worker burnout is ubiquitous, alarming, and remarkably still underreported.[22]

Reduce Manual Work and Save Time

LC/NC offspring turn up the individual automations from previous eras a notch. Teams and even entire companies can benefit as well, as we'll see throughout Part III of this book.

Simplify Transferring Content and Project Ownership

There's one final benefit of using dedicated workspaces to separate your different projects, clients, teams, and gigs: easier breakups.

Say that you're a freelance web designer working with a difficult client. You set up an independent Coda workspace for the project. A few rocky weeks follow an inauspicious start, and then things suddenly break bad. The two of you gladly decide to part ways.

If you were storing all your information in your inbox or even Google Docs, it might take some time to transfer ownership of each file and reassign permissions. Forget one, and you're likely to receive a scathing email requiring an immediate response. Transferring ownership of a workspace, however, takes just a few seconds. (Depending on the LC/NC tool you're using and your current plan, though, you may need to temporarily upgrade to facilitate that successful handoff.)

An Easy Breakup via Notion

A while back, a small business hired me to assess the quality of its internal systems and determine if an upgrade was in order. The inaugural engagement lasted for twenty hours. If we both decided to continue, we would renew after that point. From the get-go, I created a Notion workspace and faithfully kept all my documents and notes in it.

To make a long story short, the initial block of hours lapsed, and we decided to part ways. I wasn't surprised. To quote *Cool Hand Luke*, "Some men you just can't reach."

During our break-up call, my client rightfully wanted to retain access to all the documents, including my recommendations, status reports, issue logs, meeting notes, and links to our recorded Zoom calls. After all, the firm paid for the work product; its employees had every right to access that content indefinitely and use it however they wished.

In Notion, I transferred ownership of the workspace to my primary contact. All the documents remained intact. I then logged out of Notion and asked Lawrence, the IT director, to delete my access to it. If this business and I decide to work together again, Lawrence can just re-add me. Easy peasy.

Notion rocks, but I'd have followed a similar or even identical process had I used Almanac, Coda, or similar LC/NC tools.

Chapter Summary

> Contemporary LC/NC tools offer a range of benefits to organizations of all sizes. Affordability, extensibility, and ease of use are just a few.

> Allowing proper software engineers to focus on more complex development projects represents an objectively better use of their valuable time.

> Citizen developers are typically subject matter experts. They're better equipped to build certain apps than IT personnel in many cases, thus shrinking the IT/business divide.

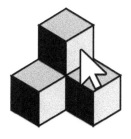

<Part III>

Unleashing LC/NC and Citizen Developers

Chapter 7

Citizen Development in Action

"The fool wonders, the wise man asks."

—BENJAMIN DISRAELI

The previous chapter described the general benefits that citizen development can confer upon individuals, teams, departments, and entire firms. Now it's time to show, not just tell.

Each of the following four case studies demonstrates how citizen developers using LC/NC tools have built powerful apps that, even a decade ago, would have required proper software developers, far heftier budgets, and longer development times.

The Municipality of Rotterdam

When many Americans think of government tech, memories of October 1, 2013, come to mind. On that day, the official site of the Affordable Care Act went live. Calling the rollout of healthcare.gov rocky would be a euphemism. Subsequent postmortems revealed the extent of its problems.[1] It wasn't pretty.

People who believe that the public sector is constitutionally incapable of getting tech right have clearly never been to the Netherlands. Visitors often remark about its government's innovative use of technology.[2] It should be no surprise then that its second largest city, Rotterdam, has embraced low-code and citizen development.

Background

In some ways, the Municipality of Rotterdam operates like most large organizations in both the public and the private sectors. It maintains nearly 1,000 IT systems to support traffic lights, an engineering department, parking tickets, public health, and other essential parts of the city's infrastructure. On any given day, its portfolio consists of roughly one hundred IT projects. (If this scenario seems familiar, it's because it resembles the one portrayed in Chapter 1.)

To churn out new systems and apps, Rotterdam historically staffed its software development department with Java and .NET experts. For the most part, these coders were only accustomed to developing and deploying applications in a linear, rigid manner. Erik van der Steen has worked in Rotterdam's IT department for over fourteen years. He notes that release delays were common, as is often the case with Waterfall-based projects. (Refer to Table 2.3 for more on this subject.)

What's more, getting these traditional developers to change their ways proved challenging. Van der Steen freely admits that the city's current software engineers were "very reluctant to start new things."[3]

The city's Information and Communications Technology department struggled to respond to its employees' and constituents' basic tech needs. ICT estimates for new projects typically ran €300,000, with wait times around one year. Needless to say, this one-two punch effectively discouraged different departments from requesting new applications. Faced with no other choices, employees understandably began searching for external development agencies to meet their needs.

Bottom line: The status quo was unsustainable, and something had to give.

Going the LC/NC Route

Van der Steen and his team knew that there had to be a better way to develop and deploy new apps. To this end, he started researching new development approaches. Agile methods—specifically Scrum—intrigued him. If done right, they would allow the city to build and deploy apps in a fraction of its previous time. Any methodology sans the right tools, however, is bound to fail. This realization led van der Steen to explore low-code/no-code, and to Mendix in particular.

Rotterdam formally launched a rapid application development (RAD) team in 2018, with van der Steen at the helm. Not long after, Mendix began to spread organically within the organization. After IT banned the use of Microsoft Access, an employee from the city's Engineering department took a Mendix course and built his first app. The results exceeded expectations. Colleagues were impressed and curious. Not long after, two more employees began experimenting. Today they're creating technical applications for their department—all the more impressive considering they're not software developers by trade.

Small, early victories paved the way for larger ones. When COVID-19 hit, the city built an app to expedite delivery relief to affected residents. The Regional Office for Self-employed Portal simplifies the process that self-employed citizens need to follow when applying for financial assistance. Because of its tight integration with the Netherlands' national-identification program, RBZ quickly verifies an applicant's identity and readily auto-populates fields such as address and identification number.

Who needs extra grief when dealing with coronavirus?

And the apps have just kept coming.

For instance, nearly 2.5 million visitors drive to Rotterdam annually. They must register their vehicles with their hosts before arriving to avoid receiving tickets. The city's previous registration solution was living on borrowed time. The RAD team worked with Mendix partner, Mansystems (now Clevr), to build the Rotterdam Bezoekers Parkeren as a hybrid app.* Within a month, 97 percent of the city's users had installed it.

RAD discovered that making continuous updates and improvements to keep up with user demands proved easier than with traditional coding. Not content to sit on its laurels, Rotterdam looked for opportunities to improve the app even more.

Another opportunity presented itself in 2020. Mendix had rewritten its core product in React Native, a popular open-source user interface (UI) software framework that ~~Facebook~~ Meta created. Although Mendix would support the hybrid app indefinitely, RAD decided to create a truly native mobile one. To this end, it contracted JAM-IT, a Ridderkerk-based Mendix partner.

A mere six weeks later, Rotterdam launched the first Mendix native mobile app in the world. The parking authorities use it to identify when vehicles with foreign license plates have repeatedly incurred fines and warrant booting. Latency is gone, users can authenticate via biometrics, and the app even works offline. Not surprisingly, city officials' satisfaction with it has skyrocketed.

---

* A progressive web app relies on several services to simulate a native mobile app.

Results and Lessons

Over the past four years, the city has developed one hundred different applications that collectively support more than 15,000 employees and 650,000 citizens. Consider the words of RAD's solution architect, Leon Schipper. Reflecting on the program's beginnings, he notes, "As a small team, we were able to grow without being formed by management." Starting from the bottom up "gave us a lot of freedom."

Note, however, the importance of reining in that freedom. Completely unencumbered citizen developers will almost certainly build apps that lack cohesion, especially with a tool as powerful and flexible as Mendix.

RAD coordinator Marja van der Veer understands this danger all too well. Ensuring a consistent design among the different apps was paramount. She describes how developing a design system and style guide allowed citizen developers to build their apps with a similar look and feel. (Chapter 10 revisits this topic.)

Back in the US, the opportunity for citizen developers in government is massive, as the following sidebar explains.

The US Public Sector Embraces Citizen Development
—Howard Langsam, CEO of OPEXUS, govtech leader
in case management software

Government operations for years have faced enormous challenges, and the pandemic and hybrid work have only intensified them. A growing percentage of government professionals retire each year with limited incoming talent from younger generations. In 2018, nearly 15 percent of federal employees were eligible to retire. By the time you're reading these words, that number will have spiked to 30 percent.[4] Despite shrinking resources, workloads continue rising. Thank a growing number of public requests, audits, and investigations.

Sadly, the technologies available to government teams have historically been terrible. Legacy systems, homegrown solutions, and outdated tools can't handle even moderate amounts of structured data. Ballooning volumes and data types make managing all those tasks downright impossible.

Something has to give.

In this crisis exists tremendous opportunity for the public sector to embrace low-code/no-code applications. In fact, this is already happening. Public-sector citizen developers are using drag-and-drop tools to automate processes and approvals previously managed in spreadsheets and email. Nontechnical employees don't require extensive IT support. Government leaders benefit from more productive offices, less strain on their IT teams, and fewer dollars allocated to expensive third-party contractors.

A Bubbly Outcome for a Budding Entrepreneur

The following case study contains pseudonyms. CitizenDev founder Vũ Trần asked me to change some names and details to comply with his nondisclosure agreement.* Note that his company used Bubble, a low-code tool discussed in Chapter 4.

Situation and Background

Predicting future demand has vexed many a sales force and sales organization. It's a freakin' tightrope. Employ too few reps, and you forgo leads, potential sales, and profits. Hire too many, and your

---

* If you watch *Better Call Saul*, you'll recognize these names and references.

bottom line suffers. To bridge this gap, many firms supplement their core sales forces through variable methods, such as outsourcing and affiliate programs.

Count Australia-based reseller HHM among this group. It sought to efficiently allocate excess leads to its affiliates in a specific way. This approach would work, but only via a custom application. Specifically, in early 2022, HHM sought to build an internal web app that could manage its leads and distribute them to its sales affiliates.

To meet this need, HHM contracted Wexler, a development shop based in the Philippines. Wexler built HHM's app in PHP, one of the fourth-generation programming languages referenced in Table 3.1. After two months, the first incarnation of the sales outsourcing management app was ready to launch. Call it SOMA1 here.

HHM employees used SOMA1 and provided positive feedback. Company founder Gene Takavic started thinking he'd inadvertently stumbled upon a new business opportunity. Sure, similar companies faced HHM's initial dilemma. Was there a potential market for SOMA? Could it be a significant source of revenue?

Takavic broached with Wexler's team the possibility of extending their initial scope of work. HHM wanted Wexler to expand SOMA for more general use.

In keeping with the lean methods that Eric Reis describes in his bestselling book *The Lean Startup*, Wexler started developing a minimum viable product of SOMA. (Let's call it SOMA2.) HHM would attempt to sell it to its partners and, if all went well, to a broader audience. Wexler agreed, and the two companies kicked off the project's second phase. Everyone waited with bated breath.

Fast-forward three months, and the initial enthusiasm had waned. SOMA2 had stalled for a couple of reasons. For starters, Wexler struggled to add new features to the existing PHP code base. It became evident that future enhancements would take a long time to deliver. Complicating matters further, Wexler didn't properly assign resources to the project. Ultimately, Takavic ended its partnership with Wexler. It just wasn't working the way that HHM needed.

Overcoming Obstacles With Low-Code/No-Code

Despite the difficulties, however, Takavic remained undeterred. He wanted to continue developing SOMA2. To this end, he began researching alternative development methods. SOMA2 needed to be easy for clients to use and for developers to tweak as needed. Any solution that required long development cycles wouldn't fly.

After some digging, Takavic landed on the concept of low-code/no-code. He eventually reached out to Vietnam-based CitizenDev, a company that builds custom desktop and mobile apps using contemporary LC/NC tools. Takavic wanted to know if CitizenDev could help him realize his vision for SOMA2.

Takavic and CitizenDev founder Vũ Trần discussed the project over several meetings. Takavic stressed the importance of low-code/no-code, but Trần raised a key point: Wexler had built SOMA2 with PHP. Using its existing code would be costly and time-consuming.

Trần and Takavic decided to rebuild SOMA2 from scratch with the popular low-code tool Bubble. Takavic loved its flexibility and the ease with which he could add the features HHM customers had requested. Trần assured Takavic that the rebuild would require a fraction of the time and money that HHM had spent with Wexler. Privately, Takavic was a tad skeptical. Based on his previous experience with Wexler, this seemed too good to be true.

The CitizenDev team began building SOMA3. Trần kept HHM informed throughout the process, and Takavic was impressed. He remarked that development was progressing quicker than he'd anticipated.

Many clients want to know *what* their partner is creating. Takavic, however, wanted to know *how* Trần was building SOMA3 so he could eventually fish for himself. To his credit, Trần was game. He let Takavic observe his team's development process. Although he lacked a background in software development, Takavic was learning general no-code principles and how to use Bubble in particular.

Outcome

Takavic was pleased to learn that, a mere one month later, SOMA3 was ready to launch. Even better, total development costs represented only one-third of Trần's initial estimate. Takavic was floored.

In the end, Takavic came to understand the inner workings of his own product. He could make a decent number of significant changes on his own. Not surprisingly, he soon found low-code/no-code religion: Based on his background, Takavic never dreamed that he would one day be able to build his own applications. HHM launched SOMA3 and an advertising campaign along with it. Sales started consistently coming in.

Future Plans

Trần and his team are currently working with Takavic and HHM on the roadmap for SOMA3 and future versions. Takavic isn't lacking ideas. Future enhancements will include task and project management. Eventually, he envisions SOMA as a niche CRM of sorts for his industry.

It turns out that Takavic's LC/NC experience is typical. "Many of my clients know a little about low-code/no-code before they contact me," Trần told me over Zoom in August 2022. "They are surprised when they see its power in action firsthand."

Low-Code/No-Code Powers Up Synergis Education

Based in Mesa, Arizona, Synergis Education works with colleges and universities to design and deliver relevant, flexible academic programs. It offers several services, one of which involves helping institutions of higher learning manage their prospective and current students. To state the obvious, these folks expect timely communications and support at every step. None of this happens today without powerful technology.

Lowell Vande Kamp serves as its CIO and chief of staff. (He's also a good friend of mine.) Synergis is one of tens of thousands of companies that relies upon Microsoft Dynamics 365 CRM, the company's popular system that incorporates both the ERP and CRM functionality discussed in Chapter 2.

Company Background

By way of background, Dynamics 365 is a SaaS offering that runs on Microsoft Azure. Regardless of which company is hosting a system, however, no mainstream one meets all its customers' needs out of the box. In this way, cloud computing and renting a system change nothing from the buy/COTS option discussed in Chapter 2.

In the past, untold numbers of organizations have attempted to overcome this limitation by doing one of the following:

> Customizing their systems by adding additional forms, fields, and database tables.
> Building integrations from their legacy systems to their current ones.
> Purchasing a third-party system and integrating it with its CRM.
> Indefinitely maintaining separate systems and applications.

None of these solutions is elegant, but storing supplemental customer, lead, and prospect information is essential. Historically, CIOs have picked whatever they deemed to be the lesser of the four evils.

Vande Kamp described to me an essential set of interactions with prospective students that Dynamics CRM failed to capture. They all revolved around what firms such as his call the *interview*. Call center reps at Synergis ask prospective students questions to qualify them. Synergis needs to accurately capture all of this recruiting data and make it available to its higher-ed clients.

For the most part, colleges and universities require prospective students to answer the same questions. As a simple example, David can't apply for a master's program if he never matriculated with his bachelor's degree. Still, subtle but essential differences exist. If Synergis can't provide its clients with timely, complete, and accurate information, they'll go elsewhere.

What to do?

The Power of Power Apps

Vande Kamp calls his team "small but mighty." It kicked the tires on the handful of best-of-breed, third-party apps on the market. They worked, but two factors deterred him from pulling the trigger. First, there was the expense of an additional software license. Beyond cost, his team would have to integrate this new system with Dynamics CRM and worry about long-term maintenance. Vande Kamp wondered if there was a cleaner, better way to solve this problem.

Enter Microsoft Power Apps, the company's main low-code/ no-code product. It lets anyone create professional-grade apps that integrate seamlessly with other products in the vast Microsoft universe. Both citizen developers and software engineers use this

LC/NC tool in interesting ways. Case in point: Microsoft reports on its website that "Toyota Motor North America employees have created more than 400 apps that help with everything from product quality control to COVID-19 screening."[5]

Intrigued, Vande Kamp held off on purchasing a third-party system for the time being. He also sent a recently hired college graduate—his daughter, Joy—to Microsoft Dynamics and Power Apps training, an investment far too few executives make. Armed with new knowledge, Joy returned to work and was excited to put it to use. It didn't take her long to create an app to track the information that Synergis call center reps gleaned from those interviews.

As its name implies, Power Apps allows citizen developers to create standalone business apps. In this case, however, the term *app* is a bit of a misnomer. Synergis did *not* create a separate, isolated app that lives outside of the company's Dynamics 365 system.

Instead, the new app appears in a new tab in the Opportunities screen of its existing CRM system. Same font, colors, and overall aesthetic. Call center reps work with a single screen and CRM system, and that's the whole point. Even better, the new fields enforce the Dynamics 365 business logic and preserve the system's data integrity. Finally, for all intents and purposes, all data exists in the same place; no more writing complex database queries.

Results

After successfully testing its new functionality in a separate environment, Synergis deployed the new Power App in its live instance of Dynamics 365—and the results have exceeded expectations.

First, issues around data quality and integrity have disappeared. Call center reps can't complete their interviews until they fill out *all* the new required fields in Joy's app. "Accountability around the interview process has also increased," Vande Kamp told me. "We now know who did what when. Finally, thanks to some fancy Power Apps logic, we can quickly tell our clients which prospective students are qualified to proceed to the next step."

In low-code/no-code, we trust.

Low-Code/No-Code Transforms a Family Business

Founded in 1987, Diamonette Party Rental is a 150-employee, family-run business based in Florida. It rents tables, chairs, tents, and cooking equipment.

Although its management had purchased and used top-quality software, spaces between departments resulted in gaps. As stopgaps, employees used spreadsheets and paper forms—hardly uncommon occurrences for small businesses.

Company president and founder Carlos Melendez was well aware of the gaps that a patchwork of Band-Aids didn't fully address. For example, employees in the tenting department struggled to find and retrieve basic information about tent inventory. The bag, size, and color of each tent were surprisingly tricky to find. One day during the early months of COVID-19, Carlos asked his son Charlie to build a database that housed information on the company's inventory of tents. When an employee entered a tent code, complete and accurate information needed to appear immediately.

Charlie wondered why his father didn't just ask for a simple spreadsheet, but Carlos knew exactly what he needed. Proper databases offer far more control over the data that users view, enter, and retrieve. (For more on this subject, see "What Specific Data Does the App Need to Store?" in Chapter 10.)

Tracking Tents

Charlie started investigating possible solutions, including Microsoft Access, Smartsheet, and FileMaker Pro. After asking a friend for advice, he landed on Airtable—a popular choice. He imported a file with tent data and, within minutes, a lightbulb went off over his head.

As a budding citizen developer, Charlie wasn't lacking ideas on how to improve inefficient and manual processes at the family business. Airtable was the match that lit the fire. The simple tent database whet his appetite. Charlie began thinking about his second project: an application that would facilitate truck inspections.

Tracking Truck Damage

By way of background, Diamonette trucks had been coming back slightly damaged a few times per month. No one knew which drivers had caused the damage because they all shared trucks. Compounding matters further, it might take a few weeks to notice relatively small dents and scratches. Determining who did what when was nearly impossible. "The drivers don't say much," Charlie says. "It's not like they're telling you, 'I hit a tree.'" Some repairs cost the company $1,000 or more.

Diamonette had affixed distinct quick response codes to each of its trucks. Figure 7.1 shows one of them.

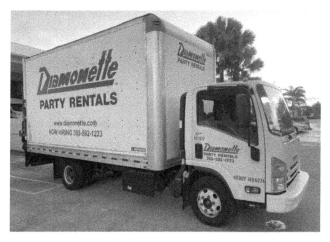

Figure 7.1: Photo of Diamonette Truck, Source: Charlie Melendez

What if the drivers used their smartphones to scan the truck's QR code before heading out on their routes? What if Charlie could systematically capture the time, date, driver name, and other relevant data? In theory, that would allow Diamonette to solve its problem.

Charlie went to work and built an Airtable Base and form that linked to each truck's QR code. After drivers scanned the truck's QR code with their phones, an Airtable form launched with prefilled information about that vehicle and conditional questions based upon its type. Drivers didn't have to waste time, enter duplicate data, and risk making errors. Charlie estimates his development time at one or two days.

The form then walked the drivers through a series of questions about cabin cleanliness, the condition of the tires, damage to the truck, and the like. Diamonette instructed all drivers to take photos after returning trucks. These photos would prove critical when management had to investigate damaged vehicles.

The Airtable app alerted Charlie immediately when Diamonette drivers reported dinged trucks, unclean cabins, and tire issues.

With a few clicks of the mouse, Charlie could search the history of each truck in the Base. For example, if Stewart reports damage on a Tuesday, Charlie can ascertain if Peter or Chris had reported it when they drove that truck on Sunday and Monday, respectively. The app made pinpointing the culprit far easier than it had been before.

Two years after launching the Airtable app, Charlie has yet to make a single change to it. In his words, it's working flawlessly. He describes it as an invaluable tool for Diamonette.

But wait, there's more.

Communicating With Drivers

Charlie was rolling and soon decided to tackle a bigger problem: communicating schedules to the company's drivers.

Nearly half of Diamonette's employees don't follow set schedules. The time of the day at which they need to arrive can vary by as much as four hours. Charlie described the legacy communication process with these eighty employees as a "nightmare." For the first thirty years of the company's existence, a scheduler would record long voicemails and expect all transportation and loading dock workers to faithfully call a designated number. Ideally, each would call the night before to learn when they needed to arrive at the company's headquarters. Think pull, not push.

Cramming eighty different schedules into a ninety-second voicemail was, to put it mildly, challenging. (Yes, that's the time limit.) Diamonette personnel needed to carefully listen to a rapid-fire message from the dispatcher every night. (He sometimes had to rerecord the message because he made an error or the voicemail had reached its 90-second maximum.) They might not have heard their names or even bothered to call, figuring that their arrival time hadn't changed for tomorrow.

One day, Diamonette trucks were backlogged. When Charlie asked the loadmaster* what had happened, he said that a single loading assistant, David, had arrived at 8 a.m. instead of his designated 6 a.m. time. That honest misunderstanding had created a ripple effect. David had just assumed he needed to show up at his usual time; he never called the number and listened to the voicemail.

In 1992, this manual communication process represented the best that Diamonette could do. Thanks to low-code/no-code, however, there was an opportunity for massive improvement. Emboldened by his previous successes, Charlie jumped on it. He wanted to build a modern app that would send drivers nightly personalized messages containing their designated reporting times.

After a week, the system was ready. Drivers and loading dock personnel now receive daily text messages with scheduling information in their preferred language, eliminating another occasional source of confusion. As of June 2022, the bespoke alert system has sent out tens of thousands of text messages to its employees without a hitch.

Results

An LC/NC tool and a curious citizen developer have automated several manual processes at Diamonette. Realizing the massive opportunity, Charlie hung his own shingle in January 2021. He works for himself as a citizen developer, specializing in Airtable. Charlie now lives in Sweden with his wife.

---

* The individual responsible for the safe loading, transport, and unloading of trucks.

I could keep going, but the length of this book needs to be reasonable. For another interesting LC/NC case study, check out how Kentucky Power used AppSheet in ways similar to those discussed in this chapter.[6]

Chapter Summary

> The city of Rotterdam launched over one hundred apps via Mendix to meet its citizens' needs.
> A small citizen-developer shop built a powerful app with Bubble that lets a client manage its outsourced salesforce.
> An edtech firm relied upon the vast array of Microsoft products to augment core systems without traditional code customization.
> A whip-smart citizen developer used Airtable to solve problems that had vexed his family business for decades.

>_Chapter 8

Approaches to Citizen Development

"Work whatever tools you may have at your command, and better tools will be found as you go along."

—NAPOLEON HILL

As the previous chapter demonstrated, one size doesn't fit all when it comes to low-code/no-code and citizen development. The fact that organizations are taking different approaches should surprise no one. After all, LC/NC tools are remarkably flexible.

This chapter takes a step back. It covers the six most common management philosophies around citizen development and their advantages and disadvantages.

Note that these philosophies may shift; they're not set in stone. New management, budgetary restrictions, security issues, and other circumstances may prompt a change.

All-In: The Single-Vendor Approach

Synergis Education from Chapter 7 follows this approach. Its vendor of choice is Microsoft throughout the firm, but it just as easily could have been Oracle or SAP.

Advantages

Dealing with a single large vendor offers several benefits.

First, consider cost and administrative simplicity. All else being equal, it's cheaper to write a single check to a sole vendor every month than ten smaller checks to ten different ones. Ditto for processing ten separate invoices. Next up are confidence and vendor stability: Microsoft isn't going anywhere.

Out of the box, the hip bone connects to the leg bone. That is, native interoperability is high. Applications and tools from a single vendor talk to each other without extensive finagling. There's less of a need for third-party tools that facilitate automation and interoperability. This level of integration is music to the ears of CIOs at mature firms who typically worry about the security issues that foreign software can introduce.

We learned in Chapter 4 that the LC/NC vendor landscape is vast and potentially confusing. Few software vendors can meet all the disparate needs of even small firms, never mind midsized and large organizations. To this end, one-stop shopping via Oracle APEX, Microsoft Power Apps, or SAP AppGyver can be appealing.

Expect fewer internal pissing matches between and among different groups with ossified tool preferences. Here's a simple example of a movie that I've seen dozens of times before:

> **Harry from HR:** We use Slack for general communication and collaboration. We found email and back-and-forth overwhelming.

Renee from Research: Interesting. We use Microsoft Teams.

Larry from Legal: We won't use either. It's email or bust for us.

Individual employees and groups won't be upset because they can't use their preferred tools on a project. Finally, garnering product support tends to be straightforward. It may be hard to find the *right* support site or number at Microsoft, but it exists. Citizen developers who use four different vendors to build their apps may find themselves stuck in purgatory.

Disadvantages

It's not all sunshine and lollipops for adherents of the single-vendor philosophy. First up, there's vendor lock-in. Switching from Microsoft Power Apps to Oracle Apex—or vice versa, for that matter—isn't going to be pretty.

A quarter-century in workplace tech has taught me one thing: Vendors typically don't make it easy for nontechnical users to extract data from their systems, applications, and databases. Moreover, no magic button converts or transforms a custom app to your new LC/NC tool. For these reasons, vendor lock-in remains a prevailing concern.[1]

Employees coming from different organizations may rue their loss of critical features in their former apps. A tight labor market means that, in extreme cases, prospects won't join your organization because they're so passionate about a particular tool.[2]

Next up, lack of vendor choice is a double-edged sword. Mice move faster than elephants. Specifically, large vendors take longer to add cutting-edge features to their wares than their smaller counterparts. The following sidebar is a case in point.

Microsoft Teams Finally Catches Up to Slack

As the pandemic continued, Zoom fatigue became a thing. People clamored for fewer synchronous meetings, but few people wanted to read 1,000-word missives.

What to do?

On September 21, 2021, Slack released Clips, a simple feature that lets users natively record and watch their colleagues' short, asynchronous videos. (In fact, the functionality existed before through Loom, other apps, and even smartphones.)

By that point, people welcomed anything that reduced the number of videoconferences and meetings they needed to attend. Users loved Clips, and the product folks at Microsoft Teams took notice. On August 10, 2022 (nearly a full year later), Microsoft announced its Clips knockoff.[3]

To be fair, Slack may be a relatively small entity, but it's hardly a mom-and-pop shop. Salesforce acquired it in July 2020.

It's not hard to find other examples of this law. I've been an Apple fanboy for over a decade, but I'm not oblivious. I recognize that the company takes its sweet time when adding features that users of non-Apple products have previously enjoyed. As but one example, iOS users can finally edit text messages—a feature that some chat-based apps have included for years.

Generally speaking, smaller software vendors release innovative features quicker than their larger brethren. Figure 8.1 shows this law in action.

Simon's Law of Software Vendors and Feature Replication

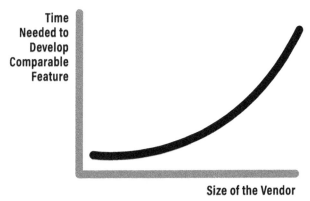

Figure 8.1: Simon's Law of Software Vendors and Feature Replication

Let's Experiment: The Skunkworks Approach

This second, self-aware approach is predicated on a simple realization: A firm's leadership doesn't know the right LC/NC tools to adopt, at least at first.

Advantages

This approach allows the organization to buy time to determine the right LC/NC tools. For whatever reason, certain ones will invariably emerge from the pack. As a result, when management formally sanctions LC/NC tools, employee commitment should be high. Leadership didn't force specific ones on its unsuspecting workforce. Besides maximizing buy-in, internecine arguments over which tool to use shouldn't happen.

Disadvantages

Try-before-you-buy may not be practical at smaller outfits; some might rightly view it as a misguided attempt to impose bureaucracy. What's more, citizen developers who began in Coda or Almanac might resent having to junk their beloved apps and rewrite them in Notion just because a VP says so.

Next, agreement from different groups is hardly a given. The decision to use Airtable might alienate the Smartsheet acolytes and vice versa. Finally, some employees may be loath to use a citizen developer's beta app that may not make the final cut.

Curated Vendors: The Best-of-Breed Approach

Against a backdrop of overwhelming choice, some organizations are whitelisting certain LC/NC tools. That is, they're selecting a handful of reputable vendors and purchasing enterprise software licenses. Writing for CIO.com, Mary Branscombe describes how some enterprises are "supporting teams or individual champions in departments and business units."[4] IT will support bespoke apps as long as citizen developers use sanctioned LC/NC tools.

Advantages

This common, deliberate, and thoughtful approach offers some juicy benefits. First, citizen developers may be able to use powerful new features faster than if their employer had relied on a large vendor's wares. (Refer to Figure 8.1.) Removing the restrictions of a single vendor introduces greater choice while concurrently minimizing the possibility of tool overload and confusion. Although no approach eliminates vendor lock-in altogether, this one lessens it. The organization isn't beholden to a single software maker.

Disadvantages

On the downside, individual employees and groups may be upset because their preferred tools didn't make the cut. It's typically more expensive to cut four checks to four different vendors than a single check to one. The gaps between different LC/NC tools may raise security concerns.

Interoperability may also suffer here. As a result, the firm may need to procure a separate license for an automation tool, such as Zapier, IFTTT, or Make. Finally, compared to SAP, Oracle, Google, and Microsoft, smaller vendors are more likely to go under or become acquisition targets.

Finally, at large organizations, this approach usually requires formal governance: establishing and enforcing policies around citizen development. Adding friction and bureaucracy can kill citizen developers' buzz and thwart momentum for their apps.

The *Laissez Faire* Approach

I described the previous strategy as *deliberate* and *thoughtful*. This next one is neither.

Advantages

A lack of guardrails and governance means that employees, groups, and departments can noodle with whatever tools they like. In theory, each entity can use its preferred tchotchkes sans interference. Employees need not wait months or years for a vendor to release a key feature; they can just freely move from one tool to the next.

In this environment, citizen developers can build whatever they want using whichever tools they like. IT departments tend to play less pronounced roles; they won't have to field as many employee support requests.

Disadvantages

This approach generally doesn't make a great deal of sense. Confusion and internal pissing matches are foregone conclusions.

Odds are that this chaotic ethos, taken to the extreme, will expose the organization to massive security issues. I'd also bet it's only a matter of time before a critical piece of tech evaporates. Finally, bean counters are likely to find inefficiencies everywhere, including several people in the company paying the same vendor-license fees.

The Wait-and-See Approach

Citizen development is undoubtedly gaining steam, but it's still shaking out. It's hardly apt to call the movement *mature*. To this end, one viable option for hidebound organizations involves sitting on the sidelines.

Advantages

At least in theory, a few years down the line, organizations can pick the most appropriate LC/NC tools. A more mature low-code/no-code market means there will be less chance of betting on the wrong horses. Citizen developers won't have to rewrite apps. Finally, if employees aren't going to build apps or employees won't use them, it's hard to justify paying for LC/NC licenses.

Disadvantages

Significant drawbacks plague this cautious approach. Specifically, consider the costs of *inaction*. For starters, employees—especially younger, tech-savvy ones—may become disaffected. Management is depriving them of tools that can make their jobs easier and satiate the new skills they're craving. (See Chapter 5.) IT will also need to field additional employee requests for new applications and enhancements to existing ones. Productivity and morale may also suffer as employee requests languish in the IT queue.

The Universal Ban

Before concluding this chapter, I'll briefly touch upon one final approach—or nonapproach, as it were. Think of it as caution on steroids.

An organization willfully and permanently forbids its workforce from using any and every LC/NC tool. Its management believes that Microsoft Office, Google Workspace, or some other sanctioned productivity suite fulfills every possible employee need. As such, it forbids people from creating apps that would help them do their jobs better. If a genuine need exists, IT should create and maintain the application or system. Period.

I, for one, wouldn't want to work in such a rigid environment—and I suspect I'm hardly alone.

Chapter Summary

> Organizations can adopt any number of different LC/NC philosophies, including none at all.
> Each approach inheres different benefits and drawbacks. Generally speaking, though, strict ones don't fly in smaller firms and startups.
> A single-vendor philosophy offers the tightest integration among different apps. On the other hand, its LC/NC tool may lack key functionality that has existed in best-of-breed counterparts for months or even years.

Chapter 9

How to Evaluate Low-Code/No-Code Tools and Learn New Ones

"The future belongs to those who prepare for it today."

—MALCOLM X

Depending on your perspective or role, the burgeoning market for LC/NC tools described in Chapter 4 is either a blessing or a curse. This chapter provides guidance on how to evaluate the ones on the market—and how to learn them once you've pulled the trigger.

Evaluating Existing Tools

You've heard good things about low-code/no-code. It easily lets users stitch together multiple components into a single mega-document or wiki. Great, but should you go with Notion, Coda, Almanac, Microsoft Loop, or something else? So many tools, so little time.

Consider the following suggestions when taking one of these cars for a test drive.

Ask Management if There's an Overriding LC/NC Philosophy

Chapter 7 profiled Synergis Education, whose CIO has gone all-in with Microsoft as a single vendor. In other words, it has adopted the first management philosophy described in Chapter 8. In this case, the choice of an LC/NC tool is simple: If Microsoft offers it, then Synergis citizen developers can consider using it. If not, then they won't or can't.

In many cases, adopting a best-of-breed approach makes the most sense. (Remember from Chapter 3 that stitching together tools from different vendors has become far easier.) If vendor monogamy isn't your employer's jam, you'll have to select from an increasing array of powerful options. Understandably, you don't want to make a mistake.

Ask Fundamental Business Questions From the Start

Einstein reportedly said that if he had an hour to solve a problem, he'd spend the first fifty-five minutes thinking about it. He'd devote the final five to potential solutions. There's no evidence that he ever uttered those words,[1] but the idea is valid and warrants a quick mention.

Any experienced website, application, database, or full-stack developer starting a new project will begin by asking fundamental questions, such as these:

> What are we trying to accomplish?
> What business problem are we attempting to solve?
> What are my constraints?

I'll borrow a phrase from Stephen Covey: They begin with the end in mind. To the extent possible, proper developers worth their salt think about their ultimate goal(s) before blindly diving into the pool or writing a single line of code. (I qualify that statement because software developers aren't omniscient. Sometimes they don't know if a particular method, framework, or library will solve the problem at hand.)

Citizen developers would do well to take a page from their counterparts' book. Foolish is the soul who believes that each LC/NC tool can do everything equally well—or at all, in some cases. You're rolling the dice if you fail to ask these foundational questions and research their answers. Six hours, days, or weeks later, you may discover the error of your ways. From the beginning, you were asking an LC/NC tool to do something that it:

> Can't currently do.
> Can't do particularly well.
> Can only do with a bunch of third-party integrations, some of which require a separate monthly license.
> Can only do with some custom code that you don't know how to write.
> Will never be able to do, no matter how many ingredients you add to the soup.

Intelligent planning and design decisions benefit any application or system, including ones built with LC/NC tools.

Money Matters

You need a new car and decide to test-drive a Lamborghini Urus. You fall in love, only to realize later that it exceeds your budget by an order of magnitude. You're disappointed, but a little research would have saved you the heartbreak.

The same principle applies to LC/NC tools.

Is Your Employer, Department, or Team Willing to Pony Up?

Sometimes the vendor's LC/NC free offering will get the job done. Perhaps the limited number of features, records, or users won't impede what you need to do.

At some point, however, you may well need to upgrade. As such, will your team, manager, department, or employer stomach the tool's monthly or annual fee? No matter how useful the software is, if the vendor's monthly or annual charge exceeds your current budget, decide whether it's even worth the time to research and test.

A team or department that typically pays for one LC/NC tool is unlikely to pony up for a closely related one. After all, ClickUp and monday.com are far more similar than dissimilar. Speaking of which …

Does My Employer Already Own a License for a Similar Tool?

Say that you love the power, flexibility, and scalability of Airtable. (Good choice, by the way.) Your company, however, writes a big monthly check for the Google Workspace suite of tools.

Google Tables is effectively an Airtable knockoff.[2] Say that it does 85 percent of what Airtable does. Are you willing to pay the extra monthly fee? Another option in this case: Continue to use Airtable under its free plan, although the IT department may bristle that you're going rogue.

Plenty of other examples exist. Notion is a remarkably flexible and useful tool, but you may find that Microsoft Loop is just as beneficial. What's more, by default, Loop plays nicely with the rest of the applications in the Microsoft 365 universe. Notion may not—at least as well.

Is the Company Behind the Tool Well Funded?

Are you about to invest significant time developing a slick business app when the vendor behind it is teetering on the brink of bankruptcy? That seems like a valuable tidbit of information to know ahead of time.

In some cases, the answer to the question is obvious. SAP, Google, Microsoft, and Amazon are loaded. They can effectively print money, but what about nearly every other vendor mentioned in Chapter 4?

Admittedly, this question can be difficult to ascertain depending on the company's age and status. (Companies in stealth mode typically aren't too chatty about the size of their coffers.) Also, tools rarely go the way of the dodo. More often than not, a big software concern gobbles it up.

Ask Tool-Specific Questions

With the general queries out of the way, it's time to drill down. While not comprehensive, here are some critical questions to ask before pulling the trigger on an LC/NC tool.

How Big and Active Is the User Community?

Although hardly perfect, proxies include the size of the tool's Facebook group(s), its Twitter followers, and the number of subscribers on its YouTube channel or subreddit. If the number is only twelve for all of them combined, the odds are that the tool isn't terribly popular. That may change down the road. In the meantime, however, understand that it hasn't exactly taken the world by storm. You may be the only person to discover a bug.

Be particularly careful of zombie groups. Say that the tool's Facebook group sports 30,000 users, but none has posted a question in the past year. Big red flag.

Does It Play Well With Other Applications, Systems, and Tools?

Perhaps you originally conceived of your app as a way to accomplish a specific task. Six months later, however, your colleagues love it, and it morphs into something else. Questions here include:

> Could you easily integrate it with other essential applications and systems?
> How hard would it be to do more?
> Will I have to move to a different, more robust LC/NC tool and rebuild the app?

Again, there's no way to know for certain if you're picking the right LC/NC tool. The notion of future-proofing an app is preposterous. In five years, you may need it to do something that's not remotely on your current radar—or anyone else's, for that matter.

Is [Insert Name of LC/NC Tool] the Right One for the Job?

To be fair, you may not know the answer to this question until you kick its tires and build new apps. At this risk of overcomplicating matters, modern LC/NC tools overlap quite a bit. There may not be a single "best" tool to do what you want.

As a general rule of thumb, though, if no one else appears to have built any close to the app that you ultimately envision, there's probably a reason:

> Your business need is unique.
> You're attempting to do the impossible.
> You've got the wrong LC/NC tool for the job.

If in doubt, ask. The r/nocode subreddit is a valuable forum for this type of thing. What's more, I've generally found LC/NC vendor support specialists very helpful.

Sign Up for the Free Plan to Kick the Tires

Once you've walked through the previous doors, start playing. Make may seem overwhelming, but compared to what? Is Zapier a more intuitive option? Google may provide an answer, but is it the right one? Compare them and decide for yourself who's right.

If several popular tools in the same category confuse you, maybe it's time to ask why. Each automation app embraces conditional logic, and you need to understand that concept regardless of which tool you select.

Try to Export Your Data

When you start using a new LC/NC tool, build a simple app. Then ask yourself how you'll be able to get the data out of it if the need presents itself. This experience will tell you a great deal about vendor lock-in. You'll have to re-create your app if you switch LC/NC tools. That's a given. Ideally, though, you won't also have to re-create the data inside of it.

Expect some degree of lock-in. Software vendors generally aren't in the business of making it easier for their clients to give them money to their competition. If, however, you simply can't export your data at all, quickly move on.

Open a Support Ticket

Fill out a help request early on. The timeliness and quality of the vendor's response will speak volumes about its support. Don't hold your breath if it hasn't acknowledged your ticket after a month. (Chapter 3 describes a boatload of community-based options for obtaining support, but even the most popular Redditors can't patch a vendor's buggy code.)

One final note here: Don't conflate users with customers. The former pay nothing. Software vendors often prioritize the latter because they keep the lights on. Some specifically reserve real-time phone support for premium customers. Again, know this going in.

Think Beyond Your Desktop

A single-user, standalone app built with an LC/NC tool can sometimes serve a critical business function for years. As a citizen developer over the past quarter-century, I've seen this happen more than a few times. Remember from Chapter 6 that the current crop of LC/NC tools can accommodate *dozens* of concurrent users. It's a quantum leap from the previous incarnation.

Citizen developers can create far beyond their own desktops. They can build apps for their colleagues, partners, clients, and others. Even the business analyst who intends to launch a single-person app for herself might wonder what happens if it's so valuable that others may eventually want in as well.

To this end, when evaluating an LC/NC tool, try to think more broadly. Questions here include:

> Will my colleagues find the tool's UI as intuitive as I do?
> Does completing a task require an inordinate number of steps?

> Do the apps I can build with this LC/NC tool require a level of technical sophistication that would inhibit widespread user adoption?

These are all excellent questions to ask. Encourage a colleague to explore the prospective LC/NC tool for half an hour. If he finds it clunky or slow, maybe the business apps that you create with it will be as well.

Learning New LC/NC Tools

Ideally, this book is inspiring you to learn new LC/NC tools and build indispensable business apps with them. Even the most skilled citizen developers, however, know that learning is a process, not a discrete outcome. They'll frequently have to do the following:

> Learn how to use new features of existing LC/NC tools.
> Stitch together different services within the same tool.
> Use automation in new ways.
> Pick up new tools altogether, especially if they switch employers.

Against that backdrop, consider the following recommendations.

Review Valuable Resources

While researching this book, I came across a geyser of information about LC/NC tools and the communities behind them. Table 9.1 displays some of the web's most valuable low-code/ no-code resources.

Site	Description	URL
No-Code Resources	A Notion page with oodles of resources.	tinyurl.com/nclcresources
No-Code Subreddit	Valuable forum for asking questions, sharing opinions, and much more.	reddit.com/r/nocode
No-Code Essentials	A hand-curated directory of the best resources about LC/NC tools. The word *insane* comes to mind. You could spend weeks here.	nocodeessentials.com
No-Code MBA	A rich repository of courses, tutorials, podcasts, and more with a seven-day free trial.	nocode.mba

Table 9.1: Valuable LC/NC Resources

Visit these sites, and, within minutes, you'll find a deluge of YouTube channels, newsletters, blogs, tutorials, podcasts, communities, and more. If you're looking for online courses, go nuts on Skillshare,[3] Coursera, and Udemy.

Experiment and Get Your Hands Dirty

Reviewing resources, reading articles, and watching videos will only get you so far. The web is a rich place, but diminishing returns will kick in. At some point, it's best to dive right in. Set up a test environment or workspace. Load dummy data. Start building, tweaking, automating, and playing. Don't be afraid to break something. Learn why it broke and how to fix it.

Pick a Lane

Polyglots are fascinating. They display an uncanny ability to learn new languages. Once you know seven different ones, picking up an eighth isn't so hard.

I can't imagine concurrently learning how to speak Chinese, French, Polish, and Arabic. Ditto for a different type of language: Ruby, Python, and Java. The same concept applies to LC/NC tools. Trying to simultaneously master seven of them is an exercise in futility. It's much easier to learn a single one and, if you like, build from there.

Dedicate the Time and Be Patient

I've seen people who are teaching themselves new tools make two types of mistakes. Each of the following approaches is unlikely to bear fruit:

> **Overly casual:** Trying to learn the new application in fifteen-minute blocks every other Sunday while on their phone.
> **Cram session:** Trying to learn everything there is to know about the app in a single six-hour block.

The research on sustainable learning in multiple fields is clear: Neither of these styles is likely to work. With respect to the first, *appropriate* spacing correlates with more effective learning.[4] Spaced review improves memory for both simple and higher-level material,[5] but the duration of the spaces matters. Take a day off if you like, but not thirteen consecutive ones. As for the second misguided approach, unless you're Rainman himself, you won't be able to pull it off. Cramming just doesn't result in long-term knowledge retention.[6]

Maybe you're still a bit skeptical. After all, how hard can it be to learn how to use a fairly limited LC/NC tool?

Let me assure you: Even one built around a primary purpose ships with more functionality than casual users realize. For example, Zapier excels at automation, but that's a robust category. If you don't believe me, spend a few minutes exploring it. You can quickly grasp the basics, but you'll discover just how much you don't know. Give it time.

Chapter Summary

> Before beginning to learn a new tool, see if there's a financial or other reason that you wouldn't be able to use it in your job.
> Assuming that you're good to go, sign up for a free plan and start experimenting. You're far more likely to learn by doing than by osmosis.
> Consistency and adequate spacing are the most important factors when teaching yourself a new tool.

Chapter 10

Successfully Navigating the LC/NC App Lifecycle

"Time is a flat circle."
—FRIEDRICH NIETZSCHE

If I've done my job correctly in the previous nine chapters, you're downright giddy to get going with low-code/no-code. Stoked, even. You can't wait to start building new apps and launching them. Wait until your peers, partners, and customers see your app and how it solves pesky problems and saves time. Maybe you're even looking to rock a citizen-developer t-shirt[*] or some other swag. If you're a manager and have previously resisted—or not known about—citizen development and LC/NC tools, you're sold.

I understand the excitement. Really. Citizen developers generally find today's LC/NC tools downright addicting. We like creating, solving problems, and streamlining inefficient business processes. Learning is our cardio.

---

[*] Yes, they exist. See https://tinyurl.com/cdswag-ps.

To borrow the title of Larry David's popular and eminently quotable show, it makes sense to curb your enthusiasm at first. Before kicking off your development and deployment efforts in earnest, it's wise to first take a step back.

This chapter serves as a framework of sorts for aspiring and current citizen developers—a gestalt in which to view their existing and future apps. No, there's no comprehensive checklist of tasks to complete. Rather, these pages present some key ideas, wisdom, and words of caution for citizen developers as they bring their new babies into the world.

I'm not reinventing the wheel here. I've loosely based this chapter on a tried-and-true framework that proper developers have used for years: the Software Development Lifecycle.[1] (Note that some people refer to SDLC as the *Systems* Development Lifecycle.) My reliance on it reflects yet another similarity between traditional programmers and citizen developers.

Planning and Gathering Requirements

The flexibility of low-code/no-code tools is both a blessing and a curse. Yes, their current options and tools offer unprecedented power. By themselves, however, those tools guarantee exactly nothing. An eager beaver can build a slick LC/NC app. For one reason or another, however, it just doesn't get the job done.

No one wants that.

Avoid that outcome by thinking about the following questions at the onset. Gathering essential requirements represents time well spent.

What Business Problem Are You Attempting to Solve?

This question resembles the one from the previous chapter on evaluating different LC/NC tools. The major difference now is that, to use a golf analogy, you've already selected the club from your bag. You now need to hit your shot.

If your colleagues, clients, and partners disagree on the problem, the odds are that your solution isn't going to satisfy their needs.

Should I Even Try to Solve It in the First Place?

LC/NC tools and their offspring can reduce friction. As we saw in Chapter 6, it's one of their primary benefits. That is, employees in different lines of business can just do it. They don't need to ask their IT departments, developers, consultants, and third-party agencies to create custom business apps for them. Power to the people, to quote John Lennon.

But is this a problem worth solving? What second-order effects will solving it create? (Yes, this existential question transcends the low-code/no-code world, but we can save that conversation for beers sometime.)

Sticking to the matter at hand, I'm all for democratizing application development and reimagining manual business processes. I wouldn't have written this book if I felt otherwise. Geeks like me die a little inside when they have to abide by inefficient ones. (Routinely copying and pasting between applications is my definition of hell. The same goes for waiting two years for IT to roll out a substandard solution—especially today.) But would automating a clunky process cause as many problems as it solves?

For instance, a fashion designer hires a dozen TikTok influencers to promote its brand for a one-month Twitter campaign. Understandably, the designer wants to seamlessly capture all the influencers' tweets in a single place. There are lots of efficient ways to accomplish this goal. One involves a slick IFTTT automation that routes all tweets into a single Google Sheet.[2] Take that, copy-and-paste.

Fantastic, but what if each influencer tweets twenty times daily—and not always about the campaign? Do the math. At the end of the month, there will be 8,640 tweets in that Google Sheet, most of which aren't relevant to the designer.

Sure, there are ways to filter those tweets. The segmentation process should take seconds if the influencers use the designated campaign hashtag. The general point remains: Just because we *can* build apps with LC/NC tools to automate a manual process doesn't mean we *should*.

Who's the Target Audience for the New Application?

If you're building an app for yourself, you can ignore the following questions. If not, pay close attention to them:

> How many people will need to use the new app? Ensure that your current plan with the LC/NC vendor supports this number. Yes, upgrading is easy, but some users may resent the early problems.

> Is using the app voluntary or compulsory? Can you compel others to use it?

> Do the users work for the same group, department, or company as you do?

> If not, will people outside your company be able to use it? Remember that plenty of organizations continue to battle shadow IT.

> Will your audience require training?

As the following sidebar demonstrates, clearly answering these questions increases the chances that the new app sticks the landing.

LC/NC Advice, Part I: Loans and Low-Code
—Steve Putman, data management expert, musician, and author

I once worked in a post-closing department for a mortgage lender. Its normal business process involved taking recently closed loan files and auditing them for completeness. Each loan file came with a six-page checklist that covered every kind of loan. Entire sections of the checklist didn't apply to certain loans.

The auditors would have to know which sections of the checklist they could ignore and which ones to cross out. At least in theory, they would fill out all remaining items. A lead auditor would then inspect that work and return loans that didn't complete the required parts of the list. The icky process reeked of inefficiency.

My boss and I created a solution that would now fall under the low-code umbrella: a small, PC-based database application* that accepted loan characteristics as user input. It then produced a simple checklist containing only the items required to audit the specific loan.

It wasn't sexy, but this little app did two crucial things. First, it took the burden of knowing the different permutations off the auditors. Second, the lead auditor no longer needed to provide a detailed description of the work because the application automatically rejected incomplete loan applications.

Boom.

After we rolled out the application, errors plummeted by more than 25 percent. What's more, since the app eliminated

---

* We used dBase III—a predecessor to Microsoft Access—and Clipper.

duplicate audits, group productivity improved. It was a big win for the entire department.

Remember that citizen developers can be effective as long as they remember the scope of the problems they're trying to solve.

Are You Building a Greenfield or Brownfield Application?

Is your app a truly new creation, or does it replace an existing one? If it supplants a current and popular app, expect a little blowback. The idea that a citizen developer created something better may rankle employees who take pride in *their* legacy apps and systems. Who the hell are you to improve *my* app?

To be fair, I've only seen this type of resentment in bureaucratic organizations from puerile employees with considerable axes to grind.

Does IT Need to Formally Sanction the App?

Let's return for a moment to the philosophies described in Chapter 8. There's no sense in trying to build an app that the IT department will block. Maybe you're not planning on asking for permission first. Fine, but sometimes employees who attempt to circumvent IT wind up in the penalty box.

Who Will Maintain the Application Once It Launches?

Citizen developers must decide if they'll give others God-level access to their creations. Assigning different admins or owners inheres additional advantages and drawbacks. A trusted colleague can take simple tasks off your plate and field support calls during your absence, but are you letting a fox guard the henhouse? Citizen developers are introducing additional risk here.

What Specific Data Does the App Need to Store?

The traditional SDLC doesn't emphasize data per se. Still, I'd be remiss if I didn't include a brief section on this critical subject. Along with automation and collaboration, two of the chief benefits of citizen development are the ability to easily gather data and preserve its quality.

All citizen developers' apps aren't created equal. There's a world of difference between the following two LC/NC creations:

> **App #1:** A simple list to track gifts for a dozen employees attending the upcoming office holiday party. Other than name and gift, this list is pretty straightforward. It's hard for even technophobes to screw up.

> **App #2:** A complex, multiuser database that lets employees manage product inventory and customer orders. Ultimately, it needs to track tens of thousands of records and dozens of fields. (Note that, in this case, low-code/no-code probably doesn't represent the best way to go. Chapter 2 described proper systems that are more suitable in this case.)

Remember that the notion of a citizen developer isn't binary; there are degrees. Ditto for the apps created with LC/NC. With App #1, a citizen developer's lack of familiarity with all things data won't inhibit app-development efforts.

App #2, however, is an entirely different ball of wax. The citizen developer who ignores critical concepts such as data normalization, data types, data modeling, primary keys, relational data, and database views will almost certainly regret it.

Josh Snow is a no-code developer based in Australia and founder of supportdept.io. He relayed to me some of the data-related mistakes that citizen developers make when they lack knowledge of this essential domain. They weren't pretty.

Risks such as data integrity arise. They can be problematic if users rely on apps for mission-critical processes and decision-making.

What should citizen developers do if they need to build a relatively sophisticated app but lack some essential technical underpinnings? To borrow from *The Matrix*, they've got two options.

First, they can take the red pill. Start doing some research or talk to a colleague who has developed databases. No, they need not become expert data modelers or database administrators. Trust me, though: That investment will pay off in spades. Some LC/NC vendors wisely recommend that citizen developers model their data *before* they begin building their apps.[3]

The blue pill is still on the table, but I'd advise against taking it. Depending on the amount and types of data that your new app will store, you need to grasp a few essentials. For example, allowing users to enter whatever information they want in a comments field makes sense. Doing so for a field with a limited number of realistic values, however, will quickly become the bane of your existence. The following sidebar demonstrates the pitfalls of allowing users free rein of data entry.

One Source, Fourteen Different Values

During my years as a college professor, I taught nine sections of the analytics capstone course.

Students completed real-world projects and helped organizations build statistical models, analyze their data, make predictions, and create interactive data visualizations with Tableau.

One summer, two groups of students worked with a local painting company looking to optimize its marketing budget. The small business had tracked all data in a Google Sheet. When my students received it, they learned a critical lesson: Data quality is frequently messy, even from small businesses.

For instance, lead source often contained eight or more disparate values for the same entity. Facebook was sometimes FB, Facebook, facebook, Fbook, and other permutations, including typos. What's more, Google Sheets can't require users to enter fields. Employees sometimes failed to include essential information, such as lead date. As a result, the first part of the project involved extensive data cleanup, at least to the extent possible. The groups couldn't make valid recommendations based on garbage data.

The students offered several suggestions at the end of the semester. The most obvious involved restricting certain fields through the Sheets data validation functionality. In fact, both groups advised the owners to quickly adopt a proper CRM system. Their professor wholeheartedly agreed.

Beyond restricting user input—or input type—for specific fields, citizen developers should think about their apps' ultimate users. Sure, they *can* store up to 500 customer, product, and order fields in a single Airtable Base.[4] Whether they *should* do it is another matter. If an app's design confounds its users, they'll likely hate it, refuse to use it, or routinely make mistakes.

If you're not concerned about the audience for your app (and you should be), then consider this: Poorly designed apps almost always result in unwieldy maintenance. Finally, you may well have to rewrite it from scratch. So, there's that …

Design and Development

What will the app look like? How will a citizen developer build it? Besides the data considerations discussed in the previous section, consider the following advice.

Think Agile

Remember that low-code/no-code apps don't supplant the need for the proper ERP, CRM, and other enterprise systems mentioned in Chapter 2. Exceptions aside, think of them more as glue apps that fulfill a valuable, specific, and possibly short-term purpose or, at most, a limited set of them.

Citizen developers should opt for an Agile development approach for this reason, save for exceptional circumstances. Their apps are likely to quickly evolve, and there's generally no need to gather all conceivable requirements from the start. The app can probably afford to go live with some features; the citizen developer can probably wait on the rest. Ideally, its users will suggest new functionality, integrations, and automations of their own—some of which never would have occurred to the citizen developer. Finally, a quick launch can create an early buzz around the app. Ditto when users see their suggestions turned into features.

Consider the Audience's Primary Device

The audience for the new app matters, but that's not the only factor that citizen developers need to consider. An app may look good on a proper computer but terrible on mobile devices, especially on smaller smartphones. Database tables with twenty-nine fields may be impossible for some people to view on their iPhones, especially if they're in their fifties or older.

Recognize the Limitations of Your LC/NC App

By definition, an NC tool provides limited ways for citizen developers to customize their applications' appearance. As a group, LC is more flexible than its NC counterpart. Regardless of your instrument of choice, no one would ever mistake even the most potent LC tool for a proper, contemporary UI development framework à la Angular or React (referenced in Chapter 7).

Consider Starting With Templates

If you're eager to launch your new low-code/no-code app, you're in luck. The rise of the tools discussed in Chapter 4 has birthed a corresponding surge in templates. Think of them as ways to jumpstart your app or automate manual tasks. At the very least, perusing them will give you ideas about what you can do and how to do it.

For example, say that you want to track your company's marketing campaign using Airtable. Start from scratch if you like, but why not check out existing templates[5] and customize them from there? Google "Notion templates" and go nuts. If you're looking to up your Bubble game, consider Atomic Fusion, a site chock-full of existing code snippets, landing pages, and even a UI kit. When I spoke with its owner Ranjit Bhinge in September 2022, he explained how these building blocks expedited app development, product launches, and idea validation for thousands of users.

At the risk of stating the obvious, people learn in different ways. I watched dozens of introductory videos when I started teaching

myself Airtable and Notion.* After a few of them, I'd sit in front of
my computer, roll up my sleeves, and start noodling. Delving into an
intricate project management template without a prior understanding
of either LC/NC tool would have helped more than it hurt.

Seek Input From Others

With rare exceptions, developers—citizen or not—think that
their apps are intuitive. They just make sense. After all, they built
them. How could any user *not* understand them?

It turns out that developers are just like the rest of us: They suffer
from a cognitive bias commonly known as *the curse of knowledge.*[6]
That is, we human beings naturally assume that everyone else knows
what we know. The curse doesn't discriminate. It afflicts all doctors,
lawyers, writers, and—yes—software developers.

To overcome it, citizen developers need to involve end users
early in the development process. The key question is, Does the app
make sense to them? Developers who ignore their audiences until
the end do so at their own peril. What good is an LC/NC app if no
one understands how to use it properly?

Design With Intent and Consistency in Mind

Log in to Facebook, Twitter, LinkedIn, or Instagram. In each
case, your feed is unique. Ditto for your name, profile pic, and
preferred language. Those differences aside, everyone views the
same user interface. That is, users can't alter the structure and
overall appearance. Unless developers are specifically conducting
an A/B test to evaluate the merits of a proposed change, everyone
will see and interact with an identical UI.

---

* Thomas J. Frank (www.thomasjfrank.com) is a productivity wunderkind.

That makes sense, right? Imagine placing your LinkedIn work experience inside the publications section. What if you put your location on Facebook underneath its photos section? The lack of standardization would confuse users. Recruiters on LinkedIn might miss your work experience and offer a lesser qualified candidate an interview. In all cases, time-on-site would eventually plunge, ad revenue would suffer, and these companies would make less money.

I'm not just speculating. Back in the day, Myspace allowed its users to significantly customize their home pages. Why not give its users the chance to infuse their personality? What could go wrong?

Hello, theory. Meet practice.

The results weren't pretty.[7] Mark Zuckerberg's creation soon overtook Myspace. The rest is history. Many execs, product heads, and developers at Twitter, LinkedIn, and scores of other companies remember the Myspace lesson well.[8]

Citizen developers need not enroll in formal design programs. At the same time, they'd benefit from understanding a few basics on the topic. As the following sidebar illustrates, a consistent overall aesthetic will likely promote the app's adoption and minimize user questions.

Why Design Matters—Even for Citizen Developers

—Matt Wade, consultant and author

The user experience aims to achieve "a deep understanding of users, what they need, what they value, their abilities, and their limitations."[9] Think of it as the intersection of digital interfaces and human psychology.

As you might expect, ~~Facebook~~ Meta, Twitter, Netflix, Apple, Google, and almost every other tech company employs proper UX specialists today. No organization wants its apps, websites, systems, and devices to confuse its users and customers. In short, design really matters.

Remember that citizen developers aren't necessarily designers of any type. Don't assume that they've all studied user experience. Yet, by using LC/NC tools to create new apps, they become *de facto* UX designers—and possibly mediocre or poor ones. The result is far too frequently an inconsistent, unintuitive, inaccessible user experience that confounds the app's colleagues, partners, and clients.

LC/NC tools run the gamut. A great deal hinges on a citizen developer's particular instrument. Some simple NC ones severely restrict citizen developers' design options. More sophisticated ones, however, let them begin with a blank canvas and go hog wild from there. They can insert radio buttons, checkboxes, logos, drop-down menus, and other elements wherever they like. Because they allow third-party code, LC tools expand citizen developers' options even more.

Left to their own devices, citizen developers within a team, department, or organization would naturally select their own colors, fonts, and icons. The resulting user experience across these different apps would be inconsistent and frustrating.

If you don't believe me, consider the following: More than thirty years after the arrival of PowerPoint, people still frequently create slides crammed with tiny text, contrasting colors, clip art, and fugly fonts. In meetings and presentations, these design atrocities result in lots of yawns and eye-rolling. Make no mistake, though: Death by PowerPoint is real. One terrible slide even helped kill seven astronauts.[10]

Organizations should provide citizen developers with templates, examples, design guidelines, and a style guide to alleviate this problem. (The Rotterdam case study in Chapter 7 illustrates the need for consistent design.) To the extent possible, take the guesswork out of the design process. Pave a path for them.

Also, take accessibility into account. Are the fonts large enough for the audience? Does your color scheme offer enough contrast? Did you provide an icon collection and guidance on when to use icons versus buttons or text? Is there an easy way to make multilingual versions? Are you pushing alt text? Have you included a high-contrast experience? And, on that note, why not enable dark mode?

Carefully thinking about these design considerations in advance won't guarantee the successful rollout of citizen developers' new apps. It will, however, prevent the citizen developer from having to bushwhack through myriad design options. Users of the app will appreciate the consistent—or, at least, less chaotic—user experience.

Provide Intuitive Support Options

How can the app's users ask questions or provide feedback? The gangster move is to bake support functionality into your app from the get-go. Include a prominently displayed support page or request form in your new Notion or Airtable creation.

Testing

A developer of any kind with a perfect track record is a unicorn. Both proper and citizen developers occasionally make mistakes and, to be sure, some are bigger than others. Depending on the specific oversight or gaffe, the citizen developer may get more than one bite at the apple. Still, why risk it? A little testing can identify early issues and maximize the app's adoption. At a bare minimum, it increases confidence that it is ready for prime time.

As a general rule, the more complex the app, the more testing that citizen developers need to conduct. Figure 10.1 displays this relationship.

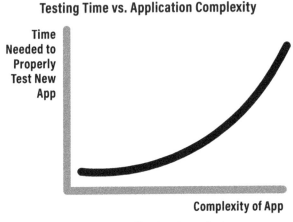

Testing Time vs. Application Complexity

Time Needed to Properly Test New App

Complexity of App

Figure 10.1: Testing Time vs. Application Complexity

Keep the following suggestions in mind. First, testing may not be essential if no one else will be using your app. Assuming that other people will, you need to test it. Even the most basic NC app benefits from a good tire kicking. It's wise to do more extensive poking around with more sophisticated LC ones. Double that if you've inserted your own code.

Remember to consider both types of errors: those of commission and those of omission. The former includes mislabeling a field in a database. Failing to include a report-sharing option falls under the second category.

Finally, just because the app works for you doesn't mean that it will work for everyone else. Have you tested it on different screen sizes and devices?* Different browsers? Operating systems?

Launch

You don't want all your development work to go to waste because of a botched deployment. Here are some ideas around initially launching your app. If you'll be using it solely for your own purposes, feel free to ignore this section.

Communication

Popular options include an email blast or posting a channel-wide message in an internal collaboration hub, such as Slack, Microsoft Teams, or Workplace by ~~Facebook~~ Meta. Newsletters and old-fashioned word of mouth work as well.

Versions and User Expectations

Besides the medium you're using to communicate the launch of your app, think about the message itself. Truth in advertising is essential. If you're launching a beta app to specifically solicit feedback and find bugs, make that clear. Users are less likely to expect your new creation to be perfect. Tell people if you're launching the app in a development or training environment. Also, you don't want people ignoring it because they're afraid it will complicate their lives.

Training

Launching an app into the wilderness can be downright exciting. Your peers, partners, or clients can immediately start benefiting

---

* Check out www.responsinator.com.

from your genius creation. Although your app's UI and features seem intuitive to you, your audience may feel differently.

To this end, seriously consider providing formal training. The citizen developer who ignores offering training for a sophisticated app is playing with fire. Figure 10.2 displays the relationship between the need for end user training and the app's complexity.

Need for End User Training vs. Application Complexity

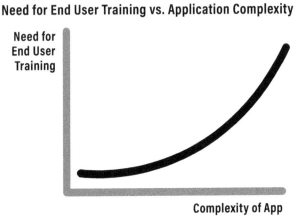

Figure 10.2: Need for End User Training vs. Application Complexity

Here are some additional questions to consider:

> Will you offer new users ongoing training?
> If yes, will it be the one-on-one or group variety?
> If synchronous learning is off the table, will you create videos, guides, a wiki, or an FAQ?
> To foster a sense of community among your app's users, will you create and monitor a dedicated channel in Slack, Microsoft Teams, or another internal collaboration hub?

When I've built apps for my clients and peers over the years, I've erred on the side of providing too much training. As the following sidebar indicates, structuring LC/NC training as a monolithic, single-purpose class is a mistake.

The Training Two-Step

—Gareth Pronovost, no-code consultant

I've built custom apps using different no-code tools for hundreds of companies. The slickest app, however, means nothing if people don't know how to use it. Whether you're a consultant or an employee, ensure that your audience receives proper training. Generally speaking, training falls into two distinct categories: solution-specific and tool-specific.

The first type is relatively narrow. It focuses on the app that my client hired me to build. This involves onboarding team members who didn't directly participate in the creation of the no-code solution. Often these folks aren't citizen developers, but they need to learn how the custom app works. People will be interacting with this new app, so they'd better learn how to use it properly. Specifically, they'll view, update, enter, and delete information.

During the second type of training, it's best to broaden the lens. Teach attendees how the LC/NC tool generally works. Cover functionality that you haven't included when building the bespoke app. Think of it as learning the tool behind the tool.

I've found that citizen developers feel especially empowered after attending. If they're itching to build their own bespoke apps, you've done your job as the instructor.

Support, Maintenance, and Documentation

Again, I assume that other people will use your app in this section.

Congratulations, citizen developer. You've built and launched a marvelous new creation that solves a thorny business issue. Your colleagues will be erecting statues in your honor and singing your praises. Thanks to your app, food will taste better. The air will be cleaner.

That's a wrap, right? To paraphrase Jay-Z, on to the next one. Think again. You're not out of the woods yet.

Support

Consider the following support-related topics:

> Will you answer users' questions? If not you, who will? IT? Other citizen developers?
> If it's just you, what happens if you go on vacay?
> How will users report issues?
> Will those users call you to report a bug? If not, should they contact Mendix, ServiceNow, Notion, or the vendor whose tools you used to create the app?
> Have you communicated the support protocol to existing users? How? What about future ones?
> Who's tracking issues, working on them, and communicating their resolution?

Better to think of these questions and communicate their answers ahead of time. Launching an app without a plan is a recipe for disaster.

Preserving Knowledge and Planning for Contingencies

At some point, a citizen developer may decide to seek greener pastures, especially during the Great Resignation. While dozens of colleagues may continue using the LC/NC app, the developer's intimate knowledge of it is unique. As the following sidebar explains, don't assume that even proper techies will immediately grasp the inner workings of a custom app.

Yeah, Yeah. We Got This ... Until We Don't

After leaving full-time HR work, I spent more than a decade helping organizations implement ERP systems. On all my long-term gigs and even many of my shorter ones, I'd invariably create Microsoft Access databases. I'd make it a point to meet with the people who'd be responsible for maintaining my creations when my engagements were coming to an end. (The sidebar "Low-Code/No-Code to the Rescue, Redux" in Chapter 11 describes how they'd sometimes live on for years after I'd departed.)

On several occasions, my clients opted to pass on a hand-off meeting. Thanks, but our intelligent IT folks will figure everything out. After all, I abided by best Access practices for naming conventions and meticulously documented my work.

Sure, but some of the database functionality was intricate. For example, a dashboard might have been linked to several forms. Some of the latter would request user input and, after receiving it, let the magic happen. More specifically, it would kick off a multistep macro that invoked queries, subqueries, file exports, and email notification.

> On several of my ensuing gigs, I received an urgent email or
> phone call. The employees who dismissed my offer to transfer
> knowledge a few weeks ago suddenly needed help—and now.
> They wanted to fix an Access report or form that worked
> correctly when I left.
>
> It turns out that sophisticated apps from citizen developers
> may not be easy to maintain and tweak after all.

Custom, app-specific documentation is undoubtedly valuable,
but citizen developers should sit down in person with their heirs
apparent. I've done this both as a full-time employee and a consultant.
Not only is it professional, but it maximizes the chance that your
soon-to-be ex-colleagues can continue to benefit from your creation.
Plus, they'll be less likely to pester you with minor questions after
you've moved on to your next gig.

Retirement

Citizen developers who remain at their employers long enough
will, at some point, probably have to retire their prized creations.
The reasons will vary, but they generally fall into two categories:
vendor-initiated and employer-initiated.

When it comes to the software vendor, common reasons include:

> > It increases its monthly fee.
> > It shuts down.
> > A shinier or cheaper new tool appears, and you want to
>   make the switch.

Management-initiated grounds include:

> > New leadership assumes control. Enough said.

> Management adopts a new, more stringent low-code/no-code philosophy. It now forbids the use of your favorite LC/NC tool and its progeny. (See Chapter 8.)
> Internal auditors disapprove of your app and put it out to pasture. Perhaps it contains sensitive customer or employee data.
> Users reject it.
> A merger or acquisition closes.
> Someone realizes that the data the app is collecting belongs in a proper ERP, CRM, or PLM system.

Here are some tips to minimize disruption and rework as the app approaches retirement.

Clearly Communicate the Timeline

On April 30, 2015, Grooveshark—a free, web-based music service—suddenly disappeared. Gone. No warning. Nothing.

It didn't take Rust Cohle or another true detective to figure out what happened: legal issues. Long-time Grooveshark users like me were surprised that the service lasted as long as it did. Free, unlimited music from 2008 until 2015. Its clock had been ticking for years.

Citizen developers—and the organizations behind them—would do well to avoid a similar outcome. As we've seen throughout this book, their apps fulfill critical business functions. When they disappear, things can get ugly. Imagine if the Diamonette drivers from Chapter 7 didn't know when to show up. No, the world wouldn't end, but clients would be displeased. Ditto for employees. Why risk irritating them, especially in a historically tight labor market?

To this end, communicate the eventual demise of the LC/NC app as soon and as clearly as possible. Daily reminders are overkill, but a one-time message two months before D-day is insufficient. Find a middle ground.

Practice Exporting the Data

Ideally, citizen developers would have considered the LC/NC tool's data-export capability before pulling the trigger and building the app. Still, things might have changed since then. Don't be surprised if the data export file:

> Arrives over email with an expiring link.
> Arrives in an unfamiliar file format, specifically JavaScript Object Notation. (JSON underpins many modern LC/NC tools. Notion is a case in point.[11])
> Arrives in an unwieldy comma-separated value format. Perhaps it contains too many records for your favorite spreadsheet program to handle. Maybe a text field contains comments with commas, complicating the importing and delimiting processes.

Better to know than not know. Don't wait until Friday at 4 p.m. to export the data when the app self-destructs in an hour.

Involve IT

Chapter 6 described how IT heads have increasingly taken a hands-off approach to citizen development. It's fair to ask if IT will assist with your app's retirement, data export, and possible migration to another LC/NC tool. Hope for the best, but plan for the worst.

Chapter Summary

> LC/NC tools make creating new apps downright exciting. Still, it behooves citizen developers to think intelligently

about the lifecycle of their apps *before* they begin building them in earnest.

> Even baller citizen developers can't create everything under the sun. Initial planning and requirements gathering may manifest that an LC/NC tool just can't solve this problem.
> Specific questions concern the app's planning, design, development, launch, support, and maintenance.
> Exporting data from an LC/NC tool may not go nearly as smoothly as you assume it will.

Chapter 11

Low-Code/No-Code and Citizen Developers: Myths and Realities

"A man's gotta know his limitations."

—CLINT EASTWOOD AS DIRTY HARRY CALLAHAN, *MAGNUM FORCE*

The past five years have seen the arrival of powerful low-code/no-code tools. Citizen developers have used them to build millions of business apps. These two terms are entering the business zeitgeist. More important than any single moniker, the trend they represent is only intensifying.

As I type these words, Amazon lists a mere nineteen physical books with the keywords "citizen developer." (Yes, that list includes the one you're reading now.) I suspect that, within a year, that number will easily exceed one hundred.

Because we're still in the first inning, there's a general lack of clarity about what LC/NC tools and citizen developers can and can't do. With that in mind, this chapter lists some of their most common myths. My intent here is to systematically dispel them.

LC/NC Tools and Their Offspring

People have dunked on *visual programming* for years. Even though the name has changed, many of skeptics' arguments against it have not. In most cases, those claims fail to hold water.

They're Just Glorified Google Docs and Sheets

Utter nonsense. If that were the case, Google wouldn't have acquired AppSheet in January 2020, as Chapter 3 discussed.

Real Developers Don't Use Them

"I want to make things as complicated as possible," said no software engineer ever.

Contrary to what many believe, proper developers use low-code/no-code for all sorts of reasons. "Even developers use tools like Notion and Airtable for cases in which a full programmatic solution would be overkill," says Jimmy Jacobson, a partner at Codingscape.[1] The more technically inclined fuse LC/NC tools with traditional coding—and the results can be astonishing.

Indian data scientist Harshil Agrawal certainly agrees. He used Airtable, Typeform, and a few other services to create an impressive expense-tracking app. It took him all of—wait for it—ten minutes.[2]

To be sure, Agrawal is more proficient in programming than most citizen developers. Still, he's no outlier in using low-code/no-code to get the job done. Consider the words of John Bratincevic, a senior analyst at Forrester. Speaking with Lucas Mearian of Computerworld in April 2022, Bratincevic says:

> While low-code is often associated with 'citizen developers,' about one-third of professional developers also use it to simplify development and speed build times.[3]

Software engineers typically rely upon low-code/no-code for scads of reasons. (They may refer to it as *scriptless*.) One everyday use is regression testing.[4] This process confirms that a recent change or enhancement to a product doesn't muck up any existing features.

Low-Code/No-Code Is Just for Small Businesses and Solopreneurs

NFW.

Take Coda, for instance. The *New York Times*, Uber, and Boston Consulting Group are just a few of the large enterprises that benefit from its software.

LC/NC Tools Are Fundamentally Insecure

If this statement is true, someone forgot to tell the IT leadership at the companies mentioned above. To be fair, the hospital clerk who stores sensitive patient information in an app built with an LC/NC tool is playing with fire. Ditto for the college professor storing certain student data. If your department forbids storing sensitive information in a spreadsheet or Google Doc, the same policy likely applies to LC/NC offspring.

Our Current Software License Covers the Vendor's LC/NC Tools

Not necessarily.

Say that you work in sales and use Salesforce as your CRM. You may think that your firm's current plan includes access to its Lightning LC/NC tools—and you'd be wrong. As of this writing, Salesforce charges a monthly user fee for access. Such is life in the SaaS world discussed in Chapter 2.

I Love X. No Other Software Vendor Will Ape Its Features. Ever.

T. S. Eliot famously said, "The immature poet imitates; the mature poet plagiarizes." He wasn't referring to Airtable, Notion, and Coda, but his words are apropos here.

For decades, software vendors have been aggressively, er, *borrowing* from each other. You can find identical functionality in comparable applications. Exhibit A: PivotTables—for my money, the most useful feature in Microsoft Excel—exist in Google Sheets and Apple Numbers. Facebook cloned hashtags from Twitter and its Stories feature from Snapchat. Reels is its lame attempt to capture some of TikTok's mojo. Outside of the commercial space, popular open-source projects frequently land on the same features—or add them after they've gained traction elsewhere.

Bottom line: You may love using a specific tool. It's folly to think that no other vendor will adopt its killer features. By the way, the converse is true: The functionality that, at present, you can only find in Zoho Creator is probably coming to Mendix. And soon.

Low-Code/No-Code Apps Will Quickly Expire

Bollocks.

In January 2022, Joanna Stern in the *Wall Street Journal* wrote a piece about the imminent demise of the third generation of wireless mobile telecommunications technology—aka, 3G.[5] In her article and the accompanying video, Stern attempts to perform several basic functions with an iPhone 4. Watch the video for yourself to see what worked and what didn't.

The main point is that all applications and systems exist within the context of a ticking clock. All code eventually expires. The only question is, how long before it does?

Surprisingly, LC/NC offspring can last for much longer than one might expect, as the following sidebar illustrates.

Low-Code/No-Code to the Rescue, Redux

In Chapter 4, I described how I built a simple Microsoft Access app for South Jersey Gas to solve a small but important payroll issue in 2004.

Fast-forward to 2010. The same organization needed to upgrade its ERP system. Its management liked the job I'd done six years before and, as it turned out, I was available.

I remembered a few friendly faces as I stepped into SJG's offices. When discussing the forthcoming system upgrade with a woman named Carol, I noticed a Microsoft Access database open on her computer. I complimented her choice of skins and fonts, as I typically chose the same ones for my own databases.

Carol laughed.

"What?" I asked.

"You don't remember, do you?" Carol responded.

I shrugged.

"You built this for us when you were here last time!"

"Does it still work?" I asked.

"Sure does. Ain't broke. Don't fix it," she chuckled.

Every Scenario Requires an LC/NC Tool and a Bespoke App

Poppycock.

Sometimes a team lead, group, department, client, or firm *should* opt for a single-purpose, traditional tool because building a bespoke app is overkill. An example will manifest why.

Consider a lightweight, straightforward, and simple project-management app such as Todoist. As I write in my last book, the team that tries to manage even a moderately involved project over email and Google Sheets is asking for trouble. In this vein, Todoist may be sufficient for the task at hand, especially if the team has successfully used it before.

Every Project or Engagement Requires an LC/NC Solution

If you only have a hammer, you think everything's a nail.

Sometimes a small team or department just needs a simple, single-purpose application for a small, six-week project—and there's nothing wrong with that. If your colleagues simply aren't interested in learning your slick new app, sticking with a familiar and previously successful one may be the way to go. Ditto if the lifecycle discussion in Chapter 10 made you dizzy.

Think of it this way: As of June 2022, 126,000 organizations used Asana for project management.[6] Teams typically require a user-friendly PM program like Asana, not an LC/NC tool that lets ambitious citizen developers create custom PM apps. What's more, attempting to replicate all of Asana's robust, native functionality just to do so with an LC/NC tool doesn't make much sense.

LC/NC Tools Can Do Only One Thing

This one's downright laughable. Two examples will illustrate my point.

First, Notion and Coda can be hard for some people to conceptualize precisely because of the opposite: They can do so many different things. Each one challenges the traditional concept of a single-purpose document or app, for that matter. Second, as its name implies, Microsoft Power Automate eliminates the need to perform manual tasks. Automation, however, is a broad category that encompasses:

> Quickly sending data between and among their favorite apps and services.
> Synchronizing files.
> Streamlining business processes.
> Configuring custom notifications.
> Scraping web data for additional manipulation and analysis.[7]

Chapter 3 covered how today's LC/NC tools differ from their predecessors. Although the former are vastly more powerful than the latter, think twice about trying to use one to build a single enterprise super app.

They Can Do Everything That Enterprise Systems Can Do

Hogwash.

Remember the discussion in Chapter 2 differentiating business applications from systems. A ten-person law firm can use Airtable to create a lightweight CRM system. However, at midsized and large organizations, even the most powerful LC/NC tools don't replace the need for robust enterprise systems. Airtable, Notion, and their ilk are remarkably useful, but they can only punch so far above their weight.

There Are No Downsides to Low-Code/No-Code

Pishposh.

As much as citizen developers enjoy using LC/NC tools to build cool apps, experienced and introspective ones will cop to the following drawbacks.

Vendor Lock-In

What if you reported to the CIO of your company? She decides that it's time to break up with Microsoft. No more Teams, Office 365, Azure, Excel, and Power Apps. The move to Slack, Google Docs, and Google Sheets might be relatively seamless, but porting over those custom business applications built on Power Apps to another LC/NC tool will take some doing.

With rare exception, low-code/no-code is notorious for vendor lock-in—and this truism applies to yours truly as well. Say that I decide to move my website off of WordPress and its current Divi theme. If I want to migrate to Ghost CMS, I'll be in for a world of hurt. The project would take weeks of my time and wouldn't be remotely fun.

Tool Overlap and Tension

For the reasons discussed in Chapter 1, the time to move many tech decisions to different business lines has arrived. Great, but that decentralization creates a new problem.

For example, say that the finance department uses monday.com for project management, but marketing relies upon ClickUp for the same purpose. The company undertakes a joint project involving the two departments.

Who's going to blink? If Finance gives in, will its employees hold a grudge that could get the project off on the wrong foot?

Lest you consider that scenario unnecessarily hypothetical, rest assured: I've seen it happen.

Multiple Tools Yield Multiple Versions of the Truth

Chapter 8 described a number of different low-code/no-code philosophies. A more decentralized approach may give individual employees, teams, and departments what it wants, but there's a downside.

In these political workplaces, employees are likely to use "their" apps to maintain "their" data. It doesn't take long before key product, employee, and customer information clashes.

Software Vendors That Reject LC/NC Will Quickly Become Irrelevant

Amazon frequently rankles its vendors and regulators through some legally questionable business practices. Sensing opportunity, Shopify has successfully adopted more partner-friendly policies. In effect, it's trying to compete by being the anti-Amazon.[8] Some ecommerce companies are zigging while others are zagging, and the same thing is happening with LC/NC.

Returning to Chapter 4 for a moment, not every software vendor labels its wares as an LC/NC solution on its website. I suspect that many of these decisions are deliberate. The low-code/no-code trend is inarguable, but plenty of companies are either bucking or ignoring it. Basecamp and Todoist currently don't paint themselves with the LC/NC brush, although both integrate well with many popular third-party services, including automation tools Make and Zapier.

Citizen Developers

Chapter 5 introduced this group. Although I like my definition of a citizen developer, it's hardly the only one. There's anything but uniform opinion about them and their characteristics.

All Are Created Equal

My friends Jason Conigliari and J. R. Camilion are nothing short of coding rockstars—and not the LC/NC kind. They were two of my five best students, and they stood above their peers on their capstone projects. They can build just about anything. I've met software engineers fifteen years their senior who can't hold a candle to either of them.

The same truism applies to citizen developers: Not all are created equal; some are just better than others. The notion of a citizen developer isn't binary. There are degrees.

They Can Do Everything That Regular Developers Can Do

LC/NC tools are more powerful than ever, but make no mistake: Compared to building apps and systems from scratch, their customizability is limited. Realize that tradeoff going in.

They Need to Be Young

Bullshit. I'm a citizen developer, and I'm not young.

Any curious, intelligent, and motivated individual can learn to build powerful business apps with today's LC/NC tools. As discussed in Chapter 5, becoming a citizen developer starts with motivation: recognizing a deficiency or problem and working to overcome it. And here's where age seems to play a role.

Everyone Wants to Become One

Some people simply lack the time and desire to learn new workplace technologies—and there's nothing necessarily wrong with that. Not everyone needs to become a citizen developer.

They Can Build Apps That Obviate Enterprise Systems

Again, not true. No matter how tech-savvy, no citizen developer will design a payroll system using drag-and-drop tools. Ditto for an intricate supply-chain management system that supports one hundred employees in fifty countries. Maybe that will change thirty years from now, but I'd bet heavily against it.

They Don't Need to Worry About Design

We saw in Chapter 10 that certain LC/NC tools give citizen developers more design freedom than others. Bubble provides a formal UI builder, but Notion provides somewhat limited functionality in this regard.

Citizen developers equipped with powerful LC/NC tools stand to benefit from a little knowledge of user experience. As for how, gaining an understanding of basic design principles doesn't require a formal degree. Bibliophiles might start with Jon Yablonski's *Laws of UX*. If reading a book isn't your jam, Google offers a certificate in UX design.[9] (Chapter 1 references the company's attempt to disrupt higher education.)

How we frame and present objects invariably affects our expectations of them. They affect our perception of everything, including how food tastes.[10] Why *wouldn't* the same tenet apply to how we feel about the software and devices we use?

A Great One Requires Zero Technical Knowledge

A pox on that claim. As mentioned in Chapter 4, a burgeoning array of LC/NC tools lets citizen developers build powerful databases and lightweight systems. Airtable, Google Tables, ClickUp, and Smartsheet are just a few available options. Within a short time, a citizen developer can begin effectively tracking customers, event attendees, or just about anything else.

Ideally, low-code/no-code apps maximize speed and minimize data redundancy. As Chapter 10 discusses, making that happen requires some knowledge about databases and related concepts.

All Their Apps Will Be Intuitive

Consider the following two examples:

1. An application begins experiencing performance and latency issues. Not unexpectedly, its users start complaining—and loudly. Support folks dutifully look under the app's hood to investigate what's happening. They discover some of the original developers' spaghetti code. To fix the issue, they'll need more than a Band-Aid. They'll have to rewrite the underlying code while preserving the app's original functionality, a process known as *refactoring.*

2. A newly hired database administrator faces a similar dilemma. She discovers that her predecessors randomly bolted new tables together. The database schema is an utter mess. The first three months of her job won't go according to plan.

The bill for poorly chosen coding methods and design decisions has finally come due in both the front-end and back-end scenarios. These issues were always present; they only manifested themselves now.

The lesson here for citizen developers may not be obvious, but it's there. No, they don't need to—and typically can't—look under the hood of the tools they're using. Software vendors take protecting their proprietary code bases seriously, not that it always works. Case in point: The popular password manager LastPass reported in August 2022 that hackers had made off with its source code.[11]

Regardless of who's doing the building, bank on certain doctrines. Poor planning, method selection, design, and execution will still plague *any* application, system, or website. No-coders, low-coders, and full-coders are all capable of creating confusing and downright awful apps. In this way, LC/NC tools resemble the integrated development environments that ship with the fourth-generation languages referenced in Table 3.1.

Citizen Developers Are Amateurs Who Can't Build Anything Useful

The examples in this book belie this claim. The general contention about amateurs is equally specious. As the following sidebar illustrates, there's a long history of essential contributions from people who lack anything resembling formal training in their fields of interest.

Bulletin Boards Spawn the Rise of the Serious Tech Amateur

—Mike Schrenk, developer, consultant, and author

Some of the world's greatest inventions stem from outsiders. For example, consider the polymath and Czech monk Gregor Mendel. In the nineteenth century, he began experimenting with sweet peas. We now know him as the father of genetics. Amateur astronomers have discovered comets, galaxies, and even planets.[12] (For more fascinating examples, check out Frans Johansson's 2004 book *The Medici Effect*.)

When it comes to the contributions of serious amateurs, though, one domain endlessly intrigues me: technology.

Tech amateurs rose to prominence in the early 1980s, buoyed by the advent of the personal computer, dial-up internet connections, and bulletin board systems.* Think of the latter as primitive, terminal-based versions of Reddit. BBSs quickly "became the primary kind of online community through the 1980s and early 1990s, before the World Wide Web arrived."[13] They still exist today, but their popularity has waned.[14] We still benefit from their impact in ways that relatively few people truly appreciate.

---

* Yes, I'm old enough to remember using them in the early 1990s.

> BBSs were seminal. They allowed like-minded, serious
> amateurs to connect, exchange ideas, and learn. (NYU pro-
> fessor Clay Shirky's excellent 2010 book *Cognitive Surplus*
> delves into their historical importance.) BBSs fostered a sense
> of community that, in turn, spawned invaluable contributions
> to protocols, code bases, and the open-source community.
> Even today, the work of serious amateurs underpins much of
> the internet's infrastructure and the software you use.
>
> Case in point: Sir Tim Berners-Lee. From 1973 to 1976,
> he attended The Queen's College, a constituent college of the
> University of Oxford, England. He received his Bachelor of
> Arts degree in physics, but his lack of a computer science
> pedigree didn't prevent him from inventing the World Wide
> Web in 1989. Queen Elizabeth II didn't care when she knighted
> him in 2004.[15]

Riddle me this: If serious amateurs can discover the field of
genetics and Uranus, why can't they develop useful apps with LC/
NC tools?

They Portend the End of Proper Developers

For two reasons, that won't happen for decades—and probably
ever.

First, LC/NC offspring work best when solving somewhat limited
problems; they're not going to serve as the backbone for multinational
conglomerates anytime soon. Second, proper programmers are the
ones creating the LC/NC tools to begin with.

If you're a software engineer, don't fret about the imminent
demise of your craft. Experienced developers know as much. A 2021

survey found that only one in six managers "expect no-code and low-code platforms to eliminate professional developers' jobs."[16]

We Should Only Trust Certified Consultants to Build New Apps

I'll close this section by addressing a longstanding business question in the context of low-code/no-code. As we know by now, these tools let citizen developers build powerful business apps. At the same time, though, it's folly to claim that any one employee wants to learn these tools and build these apps, much less all of them. (The last chapter explored some of the other common myths about LC/NC tools and citizen developers.)

What to do?

Start googling, and you'll discover an abundance of independents and dev shops for all the LC/NC tools listed in Chapter 4. (I spoke with a couple dozen of them researching this book.) As we saw in Chapter 7, these third parties can fill valuable gaps if workforces lack the time, knowledge, skills, and desire to become citizen developers. Great, but do agencies and consultants need to hold formal certifications for the LC/NC tools they'll be using?

It would be fatuous to dunk on the idea of hiring people like me to provide valuable expertise. I'd be remiss, however, not to chime in on the import of certifications. Software vendors and consultancies aren't above using certifications as cudgels to procure business from scared clients. As the following sidebar illustrates, I'm speaking from experience here.

Elements of Persuasion

In August 2000, I joined the now-defunct Lawson Software as an application consultant. Based on my background using the PeopleSoft HR and payroll suite—yes, I'm dating myself—on the client side, Lawson's management thought that I'd be able to learn its similar offering. Once I learned the software, I could teach public and private courses, consult clients, and ultimately more than justify my salary, bonus, and benefits.

They weren't wrong.

To its credit, Lawson didn't just throw promising newbies like me into the deep end to see if we could swim. Rather, I began a formal three-month certification program. I hopped on a plane and made my first voyage to my employer's Minneapolis headquarters.* I took my training seriously and passed my certification exam two weeks ahead of schedule. My manager was pleased.

I'll give myself some props for completing the program early and passing the exam on my first attempt. (I don't suck at technology. Thank you, Carnegie Mellon.) Still, I harbored no illusions: My new certification didn't equate to solving critical, real-world problems for clients under the gun to meet hard deadlines. Theory ain't practice. After ten weeks of intensive study, I didn't know my ass from my elbow, but that didn't matter. Lawson management would tout my new bona fides when pitching my consulting services.

Fast-forward to 2010. I knew the Lawson HR/Payroll suite cold. I'd logged my 10,000 hours, give or take. More than 90

---

* Fun fact: Many buildings there are connected via a skyway system. These tubes make winter runs especially interesting.

percent of the time, I'd answer my clients' questions without blinking an eye. Even some senior Lawson consultants would ask me arcane application queries in person on projects and online via different forums.

Beyond that, I'd taken it upon myself to learn the system's back end—that is, the database tables. I could sketch much of the entity relationship diagram from memory. This knowledge allowed me to create sophisticated reports that, at the risk of being immodest, few application consultants could.

On a ten-point scale, my knowledge of the suite was damn near ten. A decade earlier, it was a three. Despite the marked improvement, one restriction hampered my efforts to land gigs: I no longer worked for Lawson or one of its certified implementation partners. As an independent, I was technically uncertified. In some people's eyes, that made hiring people like me risky. To the conservative executive, hiring a certified, vendor-approved amateur represented the safer bet than bringing in an independent, uncertified expert. Understandable, but perverse.

Although Lawson no longer exists as a standalone company, the certification game is alive and well. Preferred-vendor lists make life difficult for independents. To this day, many VPs and CXOs at large firms abide by the maxim "no one ever got fired for hiring IBM."

Make no mistake: All things being equal, choosing the certified consultant over her uncertified counterpart probably makes sense. A certification by itself, however, is neither necessary nor sufficient for a successful outcome. By itself, that credential guarantees nothing. Remember those words

the next time you think about outsourcing the development
of a new app.

Chapter Summary

> There's plenty of ambiguity around LC/NC tools and what
  citizen developers accomplish with them.

> Understanding the strengths and limitations of LC/NC
  tools will help organizations, departments, and teams
  make informed purchasing, development, and deployment
  decisions.

> Citizen developers routinely use low-code/no-code to build
  amazing apps. Make no mistake, though: They're neither
  sorcerers nor full-stack engineers. More complex business
  applications and systems require the efforts of proper
  programmers.

>_Chapter 12

Playing the Long Game

> *"First they ignore you. Then they ridicule you.*
> *And then they attack you and want to burn you.*
> *And then they build monuments to you."*
> —NICHOLAS KLEIN

Employees are frustrated. IT departments and personnel can't keep up with their demands for new business applications. Citizen developers aren't waiting around. Thanks to affordable new tools, they can easily create and deploy the powerful apps their teams, departments, and entire companies so desperately need. They're automating manual processes, tracking key information, quickly disseminating relevant information, and reducing miscommunication.

What's *not* to like? Where do I sign up?

In the realm of workplace tech, I'm pretty good at reading the room at this point in my career. If researching this book has taught me anything, it's that my initial assessment of LC/NC has been spot-on. There's no way else to say it: Low-code/no-code is blowing up.

The statistics, anecdotes, and full-blown case studies in these pages support this viewpoint. Leaders and organizations are increasingly embracing the low-code/no-code ethos and citizen development. What's more, we're just getting started.

We're a far cry from unanimity, though. The overall trend may be unmistakable, but it's not hard to find LC/NC holdouts, skeptics, and deniers. I'm talking about people and entire groups who generally oppose better workplace technologies and low-code/no-code in particular.

This chapter details a few examples of the types of enemies that many current or budding citizen developers can expect to face. I then throw down the gauntlet: I challenge senior leaders to remove the obstacles inhibiting citizen developers from doing their thing. Finally, I offer management and tech tips on how to maximize citizen developers' long-term success.

The Resistance Is Real

If you're an optimist, you believe that the glass is half full. For example, consider the following statistic from Gartner introduced in Chapter 5:

> Business-led IT practices represent roughly 36 percent of IT budgets—a number that has steadily increased in the past few years.[1]

Pessimists look at that number and think, *That's it? Just 36? What about the other 64 percent? Don't they get it?*

I've peppered you with enough figures about LC/NC adoption. In this section, I'll opt for a different tact. Here are three stories of people, teams, and companies that not only failed to embrace LC/NC tools and their offspring but actively resisted them.

Trying to Fix a Broken Multiorganization Project

The names in this little yarn are pseudonyms based on the show *Succession.*

For the past five years, the large software vendor Waystar has run an annual conference. To generate some buzz, Waystar hired an outside agency, GoJo, to run an influencer program. Specifically, it would coordinate with ten people who sported large and relevant followings in Waystar's core lines of business.

I was on Waystar's radar, and GoJo's point person Gerri contacted me about participating in the program. After a little negotiating and way too many emails, I asked how we'd communicate and collaborate during the campaign. Fortunately, GoJo's employees were tech-savvy. The idea of collaborating via emails was as much an anathema to them as it is to me. (Some of them weren't even born during the heyday of email attachments.)

When I learn that my prospective clients want me to use Slack and Google Docs, I light up. I signed on for the three-week campaign. In theory, it seemed like easy money: ten total posts on different social media channels. It didn't take long, though, for my enthusiasm to wane. Soon after, things started to spiral out of control.

Campaign Tools

After our Zoom kickoff meeting, Gerri created a new Slack workspace for all GoJo employees and influencers and added everyone to a group channel. She then created separate, private channels for each influencer.

It wasn't the only way in Slack to separate individual and team communications, but the decision made sense. After all, an influencer might want to ask GoJo reps a quick group question without pestering the others. #SlackEtiquette. Multiple channels, however, multiply opportunities for misunderstandings.

Gerri also created individual, influencer-specific Google Docs and shared them with Waystar peeps. She and her colleagues inserted comments and suggestions about the influencers' proposed social media posts.

Waystar personnel also chimed in with their feedback. They understandably wanted to approve every influencer tweet, video, and LinkedIn post ahead of time. Sometimes the GoJo gang tagged me in the Google Doc; most of the time they didn't. Add in the messages in the different Slack channels and a few direct messages from GoJo peeps and, two days after the project's kickoff, I was thoroughly confused about:

> What I needed to do.
> When I needed to do it.
> Whether Waystar and GoJo employees had given me the green light.

The team was using Slack and Google Docs as project management tools. Unfortunately, neither is designed for that purpose. Even if you ignored that reality, confusion and frustration will invariably result when a team uses several applications for the same reason. It's not a matter of *if*; it's a matter of *when*.

Obtaining clarity from GoJo on whether Waystar approved my post—and if I could share it—proved to be a herculean effort. A few of the other influencers felt the same way. If it was tough for me, I could only imagine how much time Gerri and her team needed to spend collating all the influencers' posts on different networks throughout the campaign.

Pitching a Better Way

The needless back-and-forth over simple content questions continued for a few more days. At that point, I gently suggested to Gerri that we retire the Google Doc. Instead, we could use an LC/NC tool like Notion or Coda to create a bespoke app for this project—and a far better suited one. Those tools' powerful yet flexible database features would let all influencers easily track what they did, when they did it, and what tasks remained for them to complete. (You know, little things.) Due dates would alert them to pending deadlines. You just can't do that in Google Docs.

I created a new Notion workspace and mocked up a database. I then added Gerri as a guest. Figure 12.1 displays a screenshot of the prototype. It wasn't complicated.

Proposed Notion Database for Waystar Influencer Program

☰ Waystar Influencer Program

⊞ Posts   ▦ Board   +

Aa #	◎ Influenc...	≡ Content	🗓 Due Date	◎ Status	◎ Network
1	Pete	Psyched to be attending the #Waystar conference.	August 27, 2021	Posted	LinkedIn
2	Mark	#Waystar conference will be amazing.	August 27, 2021	Posted	LinkedIn
3	Lucy	Teach me a better way to work #Waystar	August 27, 2021	Posted	LinkedIn
4	Mark	The #Waystar conference starts at 1 pm. I'm stoked.	August 24, 2021	Approved	Twitter
5	Ian	Looking to learn more about modern collaboration? #Waystar conference begins shortly.	August 16, 2021	Submitted	Twitter

Figure 12.1: Proposed Notion Database for Waystar Influencer Program

Gerri thanked me for my suggestion but insisted that things were working fine and, more importantly, exactly how Waystar wanted it. (Given Waystar's whole marketing *shtick* around seamless collaboration, that irony didn't escape me.) Much to my chagrin, we stayed the course.

Things only deteriorated from there.

Consider the simple question, Can I tweet X or share Y on LinkedIn yet? You'd think that the answer wouldn't require much effort—and you'd be spectacularly wrong.

Answers required five or more discrete Slack messages and, even then, I wasn't entirely sure if I was good to go. After triple-checking, I tweeted or posted the message on LinkedIn. Several times I had to delete messages because a Wayfair employee had requested a change but never told Gerri or me. I tried to hit my deadlines, but Gerri kept moving them.

It. Was. Exhausting.

An Old-School Ghostwriting Agency Refuses to Change

Again, I've changed the names of the people and company in this story.

Wayfarer Writers is a successful Canadian ghostwriting agency with eight full-time employees. Despite its relatively small headcount, the company handles hundreds of projects in a typical year. Wayfarer's mission is simple: to turn its clients' ideas into high-quality books. In effect, it serves as a literary cupid—a matchmaker or marketplace that connects affluent executives and celebrities with experienced scribes who turn disjointed stories and ideas into proper books.

For its service, Wayfarer charges a flat 15 percent. For instance, if Wayfarer charges a CEO $100,000 to churn out a book, the ghostwriter makes $85,000. Note that all ghosts work on a contractor basis. For a typical project, anywhere from ten to thirty ghosts apply from its cadre of several hundred independent writers.

Current, Inefficient Process

Wayfarer's client-writer matching process has remained unchanged since its foundation in the early aughts. Once a client has signed papers, employees email all approved writers. These messages include project particulars, including background, timeline, requirements, and approximate fee.

Wayfarer announces new projects in a single mass email. Employees don't segment its projects into different, genre-specific blasts. For instance, there's no way for an author specializing in nonfiction to opt out of new fiction announcements. Really, there's no middle ground; ghosts receive emails with either all projects or none of them. In some cases, this means two or three extraneous messages per week. Ghosts tolerate this minor inconvenience.

Say, though, that a scribe reads a Wayfarer email blast, likes the project, and decides to apply for it. Each interested ghost responds with all requested information (in theory), including a cover letter, writing sample, book excerpt, bio, resume, links, references, and other attached documents. In practice, sometimes writers omit critical information from their applications.

Other minor nuisances occasionally plague its process:

> Ghosts' applications wind up in employees' spam folders and their submissions never make it through.
> Interested ghosts may email the Wayfarer rep assigned to a project that piques their interests. By the time the rep returns from a vacation or an unexpected absence, the client has already arranged interviews with her top-two choices. The window has effectively closed before it was really open.

Even on its best day, the matching process is rarely smooth, and it's never been transparent. Once ghosts submit their applications,

Wayfarer goes dark unless its clients request interviews. Ghosts lack visibility into their submissions' status. It's a black box. Some ghosts unknowingly make the cut to interview for lucrative gigs, but Wayfarer employees haven't provided timely updates to them. As a result, the writers unknowingly commit to lower-paid ones. A bird in the hand, right? Clients are disappointed when they miss out on the chance to work with their top choices.

For their part, Wayfarer employees are deluged with manual work. They don't acknowledge receiving each ghost's application. All in all, the process isn't ideal. After several years of experiencing it firsthand, one ghost suggested adopting a better method. This ghost also happened to be a citizen developer.

That ghost was me.

Creating a Better Experience for Everyone

I set up a fifteen-minute phone call with Marie, my primary Wayfarer contact. It was evident to me that Wayfarer could do a better job, but I knew that email wasn't the appropriate vehicle for that discussion.

Marie agreed that the current process—and the tech behind it—were dated. Manually managing ghosts' submissions by sorting through their inboxes had clearly frustrated her colleagues and her. They weren't tech-savvy, though. What could they do?

I politely suggested several LC/NC tools that would transform Wayfarer's current process and improve the experience for everyone involved. Smartsheet and ClickUp quickly came to mind, but others would have done the trick, too. Hell, even a simple Google Form with required fields would have represented a marked improvement over the status quo for Wayfarer employees.

Much like indeed.com or any popular job board, Wayfarer could publish its open ghostwriting gigs but on a private or password-protected webpage. Writers could opt to receive alerts by genre and apply for gigs with a single click. (Email blasts go *poof!*) A simple form would then walk ghosts through the application process, forcing them to complete all required fields and attach necessary documents before clicking the Submit button. Finally, a true system would allow ghosts to create proper profiles with a portfolio of go-to documents, rather than manually resubmitting them every time.

Although the ultimate call wasn't hers to make, I could tell that Marie was intrigued. She said that she'd run it up the flagpole.

Thanks, but We're Good

A few days later, Marie sent me her response. Thanks, but no thanks. Wayfarer's old-school owner would maintain the current, kludgy process. If it ain't broke, don't fix it.

A few months later, a new Wayfarer employee emailed all the ghosts on the distribution list. Unfortunately, Marie had resigned. I suspect that our conversation made her see the light.

The WordPress Revolt

Most of my clients, partners, and vendors pick up what I put down, but no one bats 1.000. Some people refuse to embrace what are far better, more efficient ways of doing things.

Matt Mullenweg knows the feeling.

At different points in this book, I've mentioned the remarkably popular content management system WordPress. (And by *mentioned*, I mean *gushed over*.) In late 2018, WordPress's Grand Poobah announced the software's forthcoming move to a new default, block-based, and visual builder named Gutenberg.

Now, Mullenweg is a thoughtful dude. His rationale was solid— and certainly in keeping with the software trends described in this book. He argued that Gutenberg ships with five to ten times more functionality than the classic editor contained. In his words:

> ... there [are] more buttons; there [are] more blocks. That is part of the idea—to open up people's flexibility and creativity to do things they would either need code or a crazy theme to do in the past. And now we're going to open that up to do WordPress' mission, which is to democratize publishing and make it accessible to everyone.[2]

Sounds exciting, right?

Unfortunately, a sizeable portion of the WordPress community lacked Mullenweg's enthusiasm—and they weren't afraid to voice their displeasure. Google "WordPress Gutenberg sucks" and start parsing through the 453,000 results.

Not long after the announcement, WordPress developers created a new plug-in that restored its erstwhile, text-based editor. Users downloaded it 600,000 times after it launched.[3] (Four years later, that number exceeded five million.) Mullenweg even conceded that software specifically designed to undo Gutenberg represented the fastest growing plug-in in the storied history of WordPress.

In this important way, low-code/no-code shares a great deal in common with traditional systems, applications, and software-development frameworks. It's silly to think that everyone will always accept major changes to them. In fact, sometimes the opposite happens.

This begs the question, What can we do about it?

Management Strategies

No tech exists in isolation—and low-code/no-code is no exception to this rule. It's time to don my management hat and offer some advice on citizen development.

Expect Some Resistance to Citizen Development

Chapter 6 described the manifold benefits of low-code/no-code. It stands to reason that your boss and colleagues will embrace the apps you create with them.

Maybe, but maybe not.

Perhaps the three examples at the start of this chapter didn't resonate with you. Maybe you work for a government agency. Management not only fails to approve the use of a specific LC/NC tool, but explicitly forbids it. Ditto for any app that you may create with it in the future. Even if the top brass ultimately sanctions the application, there's no guarantee that your colleagues will use it.

To be fair, this second point may not matter. Perhaps you're the only person in your department who uses a slick Coda dashboard for tracking prospects. Remember from Chapter 5, however, that apps built with contemporary LC/NC tools facilitate collaboration far more than their predecessors ever did. Ideally, you're not just creating your own personal mini app.

Finally, the trend toward accepting citizen development is just that: a *trend*. Plenty of senior executives are still fighting it tooth and nail.

Recognize That Low-Code/No-Code Is Here to Stay

Work in tech long enough, and you'll eventually encounter an entrepreneur or senior leader enamored with a sexy new toy. Yes, I'm talking about shiny object syndrome.

A hypothetical but common scenario involves Jesse, the new CIO of a large organization who falls in love with a new system, methodology, or tech. He immediately summons his team to investigate it. They do, only to realize that he has lost interest over the past few weeks. Jesse's IT directors soon begin ignoring their boss's instructions to investigate the latest and greatest tech. They know he'll move on to something else in a month or so.

Yes, low-code/no-code has evolved from visual programming and previous generations of programming languages. (Chapter 3 taught us as much.) It's absurd, however, to claim that the category is a passing fad. Shiny object syndrome doesn't apply here.

Invest in Employee Training

We saw in Chapter 7 how Synergis Education used Microsoft Power Apps to tweak Dynamics 365. That wouldn't have happened if CIO Lowell Vande Kamp didn't send a new employee to training classes to learn how to use those new technologies.

I've made this point in most of my recent books, but it's equally important in this context: Don't expect employees to learn the ins and outs of new tech on their own time and dime. The investment in employee training is a worthwhile one. Apart from making employees more productive, it signals to them that their employer and manager value them. Planting that seed doesn't hurt, especially in a tight labor market.

Publicize Successes

Go to the website of any LC/NC vendor discussed in this book. You'll quickly find effusive tales of how [insert name of tool] helped a company, department, team, or individual do something cool. Maybe there's even an ROI calculation, although it's best to take it with a fifty-pound bag of salt.

No one is judging here. I'd probably do the same thing if I ran marketing for a large software vendor. Customer success stories are the quintessential hygiene factor. Including them may not mean all that much to prospects. *Failing* to include them, however, will probably raise suspicions. In this case, a would-be customer may well visit another vendor's site.

Promote internal victories, solutions, and improvements from LC/NC offspring at your company. These powerful stories can create critical momentum within an organization, department, or team. (Self-aggrandizing case studies from vendors won't pack the same punch.) Employees start to ask if there's a better way for them to do something. "Hold on a second. We can automate *that*?"

Let the buzz build from there.

Remove the Shackles

Chapter 1 described the years-long war that many IT departments have conducted against shadow IT. For the most part, today IT has waved the red flag.

Organizations are realizing that the benefits of business-led IT app development far exceed their drawbacks. That is, they're increasingly welcoming technologies built or bought by business lines, while providing governance and vendor management support.

Tech Strategies

You'll get no argument from me about the manifold benefits of low-code/no-code. At the same time, firms, departments, formal and informal groups, and individuals regularly misuse them in a bunch of different ways. Here are some tech-related tips to maximize success.

Think Breadth, Not Depth

Carina Sorrentino is the head of customer content at Mendix, one of the LC/NC vendors discussed in this book. "Most of our clients develop dozens of custom apps," she told me. "In the case of large enterprises, there might be hundreds of them."

To state the obvious, each of these apps isn't terribly broad. That is, a single creation doesn't attempt to track every possible type of enterprise information. Rather, each solves a specific and fairly narrow problem.

Consider the Business Need

Depending on the type of organization, it will need to effectively use several essential technologies. Returning to Chapter 2 for a moment, consider the following types of enterprise systems:

> Customer relationship management
> Enterprise resource planning
> Learning management system
> Product lifecycle management
> Ticket management

A large firm in the transportation, construction, aerospace, defense, or life-sciences industry without a proper PLM will have to tap out. Ditto for even a small college trying to make do without a proper LMS. Good luck using email and Google Sheets to grade student exams and accept their assignments.

In these cases, low-code/no-code simply isn't the answer. As a general rule, the larger the entity and the more complex the business problem involved, the more a proper enterprise system represents the best solution. (I use the term *entity* deliberately. A small team within a large department or organization may find it best to build its own app with LC/NC tools. In fact, this type of thing happens daily.) Figure 12.2 displays a visual representing this general rule.

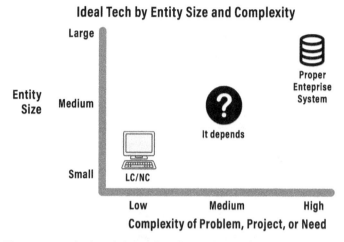

Figure 12.2: Ideal Tech by Entity Size and Complexity

Before continuing, a few disclaimers about Figure 12.2 are in order. First, think of it more as a general construct, not a hard-and-fast rule. Second, as Chapter 11 covers, LC/NC isn't the sole purview of small businesses and solopreneurs. Far from it. Case in point: Eighty firms in the Fortune 100 use Airtable, and half of those in the Fortune 1000 do.[4] Nary a single one uses it to replace a CRM, ERP, or another one of the enterprise systems listed at the start of this section.

Next, be careful with numbers, especially regarding team, department, and company size. Consider the following two scenarios:

> **Scenario A:** A citizen developer at a boutique consulting outfit builds a quick project-management app. Each of the firm's dozen consultants uses it.

> **Scenario B:** A precocious new Accenture hire in the health-
care vertical creates a kick-ass app for fifteen consultants
to address an ad hoc need.

Yes, a full 100 percent of the 12 employees in the first scenario can
use the app. Does that mean that all 710,000 Accenture employees
can do the same?[5]

Not a chance. Size matters.

Finally, when assessing the right tool for the job, the two extremes
in Figure 12.2 are easy calls to make. As with life, judgment is
essential in murky situations.

LC/NC Advice, Part II: Scope Matters

—Steve Putman, data management expert, musician, and author

I've worn quite a few hats in an IT career spanning more
than three decades. I was a citizen developer in the early 1990s,
long before anyone had heard the term. My boss and I looked
for small, tactical issues that standalone applications could
solve. We knew that the larger IT organization would never
adopt our solutions, but so what? We didn't have to follow
any corporate-wide development standards or interfaces. We
could fly under the radar.

I relate this tale to caution you about these types of glue
applications: If you're thinking about anything larger than a
departmental solution in a large company, you need to consider many other factors. Examples include tool and coding
standards, data interfaces, security and compliance, and
maintenance planning. You stand a better chance of success
if you keep your app small and tactical.

In my experience, and contrary to popular belief, IT departments generally approve of these smaller efforts because they reduce some of their system development. However, these solutions become irritants when they do the following:

- Manipulate standardized data without providing audit trails.
- Publish results to the public.
- Grow beyond departmental boundaries.

Actively Seek Appropriate Opportunities to Use LC/NC Tools

"Because that's the way we've always done it."

I'm not a fan of that aphorism, especially when I hear it after asking why a team, department, or company follows an excessively manual process that routinely confuses everyone involved.

I prefer Homer Simpson's famous if ironic quip, "Self-improvement has always been a passion of mine." The idea that we've landed on the permanent and perfect tech stack—as many techies call it—is absurd. Ideally, you're evaluating your projects, interactions, and challenges every few years. Occasionally, it's wise to kick the tires on new tools.

The following sidebar represents a simple example of how an LC/NC tool improved a process that already involved contemporary collaboration technologies.

A Novel Notion

Three years ago, I ghostwrote a book with Greg, the CEO of a software company. We expected the project to last about six months. Due to differences in time zones and schedules, we'd do most of the work asynchronously via collaborative tools, including:

- Calendly for scheduling.
- Trello for project management.
- Zoom for video calls.
- Google Docs to easily share written content and comment on it.
- Slack for general communication, research, and file sharing.[6]
- Loom to record quick, asynchronous videos.*

Although Greg was certainly tech-savvy, he found the sheer number of tools overwhelming. I understood his consternation, but each one served a unique and necessary purpose.

To be fair, I was walking the talk. That is, I was using all those applications—and more, for that matter. I was writing the manuscript in Microsoft Word and storing versions in Dropbox. Grammarly helps people write real good. (Yes, I'm joking.) For creating figures, Canva is the bee's knees for amateur designers. Otter.ai transcribes interviews. You get my drift.

Consolidating Tools

I reduced the size of my toolbox on subsequent ghostwriting projects. (Thanks for the inspiration, Homer.) Simplification,

---

* Slack Clips now largely performs the same purpose.

baby. With the concept behind Figure 12.2 firmly in mind, I went the LC/NC route.

Specifically, Notion replaced Trello, Google Docs, and even my beloved Slack to a certain extent. I created a simple Notion dashboard with links to Calendly, Zoom, and a few other relevant tools. One-stop shopping at its finest.

As a result, my clients and I now spend less time toggling among different apps. On my most recent projects, we store all documents, podcasts, articles, videos, content, and research in Notion. To paraphrase Tiago Forte's recent book, it has become our second brain.*

Notion is now one of my go-to apps, and not just for ghost-writing projects. I use it extensively for consulting, speaking, writing, and research purposes. It's even my lightweight CRM system.

Minimize Tool Overlap and Redundancy

Returning to Chapter 8, it's best to avoid using four separate LC/NC tools in your organization to perform the same functions.

For starters, a multivendor approach typically costs more than its single-vendor counterpart. Beyond that, it will increase the chances that different teams, departments, and divisions will dig their heels in on new collaborations and projects; they'll insist that everyone uses *their* preferred tools. Finally, interoperability and extensibility only go so far. It's folly to think that you'll always

---

* Forte appeared on my podcast in June 2022 to promote his new book. Listen to it at https://tinyurl.com/tiago-phil. We recorded over Zoom, and I kept thinking that I was talking to his doppelgänger Elon Musk. Dude looks just like him. It's scary.

be able to stitch together all functionality and data between two disparate applications, even with the involvement of IT.

Admittedly, context matters. Against a backdrop of decentralized IT, business-driven purchasing decisions, and citizen development, standardization throughout the organization is a tall order indeed. Rather than scolding employees for using disparate and unapproved LC/NC tools, it's best to emphasize the benefits of relying on common ones. Think carrot more than stick.

The team that uses ClickUp, Trello, Notion, *and* Coda to manage the same project is asking for trouble. All things being equal, the fewer the number of LC/NC tools used on a project, the less the chance for conflict. Figure 12.3 provides a quick visual of this key principle.

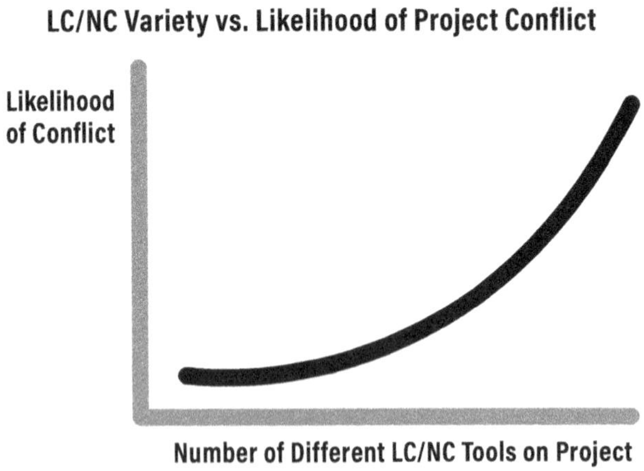

LC/NC Variety vs. Likelihood of Project Conflict

Figure 12.3: LC/NC Variety vs. Likelihood of Project Conflict

Automate, Integrate, and Do More

Using Airtable, you can quickly create a marginally better Google Sheet. If that's all you need Airtable to do, have at it. There's no sense in unnecessarily complicating a simple task, at least for now. (Whether you'll soon have to expand your tracking tool in a way that Google Sheets simply can't accommodate, however, is another matter.)

On the other hand, low-code/no-code really shines when citizen developers add intelligent automation. Explore which manual processes you can automate. Exploit the wow factor. Look for ways to reduce or eliminate redundant sources of information and duplicate data entry. Stitch your new LC/NC creation with other essential business apps and systems when desirable.

Will this mindset convert all LC/NC haters and colleagues wedded to manual, inefficient, and downright maddening ways of working? Maybe not, but rational people will find it increasingly difficult to ignore or dismiss these powerful, user-friendly, integrated, and device-agnostic business apps. (That's why I quoted Nicholas Klein at the start of this chapter. #bookends.)

Chapter Summary

> Don't expect everyone in your group, department, or company to embrace citizen developers. Like it or not, some people may well view them as threats.
> The greater the complexity of the problem, the less it makes sense to build an app with an LC/NC tool.
> A combination of sage management and tech strategies can maximize the impact of LC/NC tools and citizen developers.

Where Do We Go From Here?

"The future is already here; it's just not evenly distributed."

—WILLIAM GIBSON

Pick a contemporary technology. Some people, teams, departments, companies, industries, and even countries hop on board faster than others, *but they all eventually get there.* Whether it's smartphones, mobile payments, cloud computing, ERP, or something else, the history of technology has taught us as much.

I'll put a bow on the book by offering a few words about the future of low-code/no-code, citizen developers, and what comes next. Yep, it's prediction time.

Low-Code/No-Code Grows in Scope and Sophistication

As we saw in Chapter 4, software vendors of all sizes are wisely leaning into the LC/NC movement. Senior leaders at these companies clearly understand the importance of citizen development. To this end, they're building more LC/NC functionality into their wares and making strategic acquisitions. (The likelihood of a

coming recession means many startups will face a down round.
Their investors might decide to cash out.) Salesforce, Microsoft,
Amazon, IBM, SAP, and others might be able to acquire valuable
and complementary pieces on the cheap.

That's not to say that the smaller vendors are standing pat.
They're not, and two recent events warrant mention here. First, in
September 2021, Notion gobbled up Automate.io.[1] Expect Notion
to increasingly integrate with third-party apps and services.

The argument that Notion is just a mildly improved version of
Google Docs is already silly; in the coming months, it will become
downright preposterous. Notion's competition will have to up its
game. Ultimately, the citizen development movement benefits.

Second, consider Okta. If the name doesn't ring a bell, more
than 14,000 customers currently license its powerful authentication
software. Single sign-on isn't the sexiest concept, but it's a critical
one. Imagine the alternative: IT departments manually maintaining
dozens of different usernames and passwords for each app, service,
and system that each employee uses in a given week.

You might be asking yourself, how bad could that possibly be?

For context, here's a remarkable statistic. In March 2021, Okta
reported that the average customer relied on eighty-eight different
business applications.[2] With some companies, that number was
twice as high.

Let that sink in for a moment.

To overcome this challenge, IT administrators have historically
created and run complicated scripts that add new hires to these
different systems and services. If this sounds like the very type of
problem that low-code/no-code and citizen developers can solve,
you've been paying attention while reading this book.

In November 2021, the company introduced Okta Workflows,
an LC/NC tool that augments the company's core offering. In

September 2022, I spoke with Max Katz, the community lead for Okta Workflows. Over Zoom, he explained to me how IT administrators no longer need to write and maintain these scripts. In short, they can make their lives much easier. Automations trigger when new employees join an organization. Within minutes, employees can access essential systems and applications. Even better, the Okta Workflows product integrates with a range of third-party apps and services.[3]

Smarter Software

Advances in machine learning, artificial intelligence, and robotic process automation will supercharge existing LC/NC tools. (Geeks like me can't wait.) Applications and systems are already effectively learning how their users interact with them. Some even nudge users with specific recommendations about helpful features and integrations of which they're probably not aware. WalkMe and Pendo are just of the two newer vendors making inroads here. The latter claims to create "software that makes your software better."[4]

The Next Frontier in Business Applications?

Every day, millions of people talk to Apple's Siri, Amazon's Alexa, Microsoft's Cortana, and other virtual assistants to perform any number of tasks. We have long passed the days of only being able to say, "Hey, Google. What's the weather today?"*

Is it inconceivable that, at some point, we'll be able to build apps not with our hands, but with our mouths? Is programming by voice the next frontier in software development.[5]

Maybe.

---

* The first time that I tried Google Home, I asked it, "Who is Phil Simon?" Given that other people share my name, I was pleased that it spit out the correct answer, at least to me.

Startup Serenade has created AI-based software that lets developers speak their code.* In late 2020, the company raised $2.1 million in venture funding to expand its offering.[6]

Now, let's not overdo it. Developers won't be recycling their keyboards anytime soon. At the same time, it's not hard to envision massive improvements in voice-based programming in the coming years.

Consider the following parallel.

In 1998, say that you made the following three predictions. First, the primitive visual programming applications of the day would morph into powerful, affordable, extensible, and useful tools by 2023. These LC/NC tools would, in many cases, make software engineers superfluous and democratize app development. Citizen developers would be able to build valuable business applications that solve the types of problems discussed in this book. Finally, senior management and IT departments wouldn't just allow this practice inside their organizations; they'd actively encourage it.

Pretty crazy, right?

Parting Words

I can't tell you if voice-based programming, self-learning software, the metaverse, or other trends will fizzle. Which low-code/no-code vendors will emerge victorious? No idea. I'm not that smart.

I do know, however, that the future is bright for LC/NC tools and citizen developers. Jason Wong of Gartner echoes this sentiment. He notes that, "Business units are also increasingly controlling their own application development efforts, of which citizen development will play a crucial role in the future of apps.[7]

---

* Watch it in action at https://tinyurl.com/ps-serenade.

All signs point toward the trends described throughout this book intensifying for the foreseeable future. The LC/NC and citizen development movements are exciting and, best of all, they're just getting started.

To quote Donnie from *The Big Lebowksi*, "Mark it, Dude."

>_Thank-You

"To defend what you've written is a sign that you are alive."
—WILLIAM ZINSSER

Thank you for buying *Low-Code/No-Code*. I hope that you've enjoyed the preceding pages. Ideally, you've found this book informative, and it has challenged your assumptions about workplace tech. Its advice and knowledge will help you achieve your professional goals. And perhaps you're willing to help me.

Doing each of these things is helpful:

> Writing a book review on Amazon, bn.com, GoodReads, or your blog. The more honest, the better.
> Mentioning this book on Facebook, Reddit, Twitter, Quora, LinkedIn, and other sites you frequent.
> Recommending it to anyone who might find it interesting.
> Giving it as a gift.
> Checking out my other titles at www.philsimon.com/books.

I write books for several reasons. First, that's just what a writer does. Second, I believe I have something meaningful to say about an important topic. Third, I like writing, editing, crafting a cover, and everything else that goes into the puzzle of writing a good book. To paraphrase the title of an album by Geddy Lee, it's my favorite headache.

Fourth, although Kindles, Nooks, and iPads are downright cool, I enjoy holding a physical copy of one of my titles in my hands. In our digital world, creating something tangible from scratch is glorious. Fifth, I find writing to be incredibly cathartic. Finally, new books open professional doors for me.

At the same time, producing quality text is no small feat. Every additional copy sold helps make the next one possible.

Let me know if I can help your company, department, or team.

Phil Simon
www.philsimon.com
November 20, 2022

>_Acknowledgments

For making this book happen: Luke Fletcher, Karen Davis, Jessica Angerstein, Vinnie Kinsella, and Johnna VanHoose Dinse.

For their contributions: John Elder, Matt Wade, Steve Putman, Lowell Vande Kamp, Carina Sorrentino, Vũ Trần, Mike Schrenk, Gareth Pronovost, Josh Snow, Ranjit Bhinge, Howard Langsam, Elizabeth Orzechowski, and Max Katz.

A tip of the hat to the people who keep me grounded and listen to my rants: Dalton Cervo, Rob Hornyak, Hina Arora, Daniel Teachey, Eric Johnson, Emily Freeman, Chris Olsen, Steve Katz, Bruce Webster, Michael Viola, Joe Mirza, Dave Sandberg, Chris McGee, Scott Berkun, Josh Bernoff, Karin Reed, Martin Traub-Warner, Andrew Botwin, Moneet Singh, John Andrewski, Jennifer Zito, Rob Metting, Prescott Perez-Fox, Monica Meehan, Jason Horowitz, Marc Paolella, Peter and Hope Simon, Adom Asadourian, Helen Thompson, Kevin Daly, Sarah Garcia, Jason Conigliari, J. R. Camilon, Mark Pardy, Dustin Schott, Daniel Green, Matt Wagner, Michelle Gitlitz, Brian and Heather Morgan, Laurie Feuerstein, and especially my *consigliere*, Alan Simon.

Nietzsche once said, "Without music, life would be a mistake." He wasn't wrong.

For decades of incredible music, thank you to the members of Rush (Geddy, Alex, and Neil) and Marillion (h, Steve, Ian, Mark, and Pete). Your songs continue to inspire millions of discerning fans. I'm proud to call myself one of them.

Vince Gilligan, Peter Gould, Bryan Cranston, Aaron Paul, Dean Norris, Anna Gunn, Bob Odenkirk, Betsy Brandt, Jonathan Banks, Giancarlo Esposito, RJ Mitte, Michael Mando, Rhea Seehorn, Michael

McKean, Patrick Fabian, Tony Dalton, and the rest of the *Breaking Bad* and *Better Call Saul* teams have inspired me to do great work. Thank you for a remarkable fourteen-year run.

Finally, to my family: Thank you.

>_Bibliography

Brooks, Frederick. *The Mythical Man-Month: Essays on Software Engineering, Anniversary Edition* (2nd Ed.). Boston: Addison Wesley, 1995.

Carr, Nicholas G. *The Big Switch: Rewiring the World, from Edison to Google*. New York: W. W. Norton & Company, 2013.

Christensen, Clayton M. *The Innovator's Dilemma: When New Technologies Cause Great Firms to Fail*. Boston: Harvard Business Review Press, 1997.

Cohen, Adam. *The Perfect Store: Inside eBay*. Boston: Back Bay Books, 2003.

Forte, Tiago. *Building a Second Brain: A Proven Method to Organize Your Digital Life and Unlock Your Creative Potential*. New York: Atria Books, 2022.

Johansson, Frans. *The Medici Effect: Breakthrough Insights at the Intersection of Ideas, Concepts, and Cultures*. Boston: Harvard Business School Press, 2004.

Kim, Gene, Kevin Behr, and George Spafford. *The Phoenix Project: A Novel about IT, DevOps, and Helping Your Business Win*. Portland, OR: It Revolution Press, 2018.

Krug, Steve. *Don't Make Me Think: A Common Sense Approach to Web Usability*. Indianapolis: Que, 2000.

Project Management Institute. *Citizen Development: The Handbook for Creators and Change Makers*. Newtown Square, Pennsylvania: Project Management Institute, Inc., 2021.

Rico, David F., Hasan Sayani, and Saya Sone. *The Business Value of Agile Software Methods*. Boca Raton, FL: J Ross Publishing, 2009.

Ries, Eric. *The Lean Startup: How Today's Entrepreneurs Use Continuous Innovation to Create Radically Successful Businesses*. New York: Currency, 2011.

Shirky, Clay. *Cognitive Surplus: How Technology Makes Consumers into Collaborators*. New York: Penguin, 2010.

Shirky, Clay. *Here Comes Everybody: The Power of Organizing without Organizations*. New York: Penguin, 2009.

Simon, Phil. *Project Management in the Hybrid Workplace*. Gilbert, AZ: Racket Publishing, 2022.

Simon, Phil. *Reimagining Collaboration: Slack, Microsoft Teams, Zoom, and the Post-COVID World of Work*. Gilbert, AZ: Motion Publishing, 2021.

Simon, Phil. *Too Big to Ignore: The Business Case for Big Data*. Hoboken, New Jersey: Wiley. 2013.

Sims, Chris, and Hillary Louise Johnson. *The Elements of Scrum*. Foster City, CA: Dymaxicon, 2011.

Stone, Brad. *The Everything Store: Jeff Bezos and the Age of Amazon*. New York: Back Bay Books/Little, Brown and Company. 2013.

Thompson, Clive. *Coders: The Making of a New Tribe and the Remaking of the World*. New York: Penguin Press, 2019.

Vance, Ashlee. *Elon Musk: Tesla, SpaceX, and the Quest for a Fantastic Future*. New York: Ecco, An Imprint of HarperCollins, 2015.

Watts, Duncan J. *Everything Is Obvious: Once You Know the Answer*. New York: Crown Business, 2011.

Yablonski, Jon. *Laws of UX: Design Principles for Persuasive and Ethical Products.*: Sebastopol, California: O'Reilly Media, 2020.

>_About the Author

Phil Simon is a dynamic keynote speaker and a world-renowned authority on workplace technology. He is the award-winning author of a dozen previous business books, most recently

Thalacker Photography

Project Management in the Hybrid Workplace. He helps organizations communicate, collaborate, and use technology better. *Harvard Business Review, MIT Sloan Management Review, Wired*, NBC, CNBC, *Bloomberg BusinessWeek*, and the *New York Times* have featured his contributions. He also hosts the podcast *Conversations About Collaboration*. Simon holds degrees from Carnegie Mellon and Cornell University.

>_ Index

>_Endnotes

>_Chapter 1

1. PricewaterhouseCoopers. 2018. "Our Status with Tech at Work: It's Complicated." October 2018. https://tinyurl.com/529axcyw.
2. Katie Costello. 2020. "Gartner Says Worldwide IT Spending to Grow 4% in 2021." Gartner. October 20, 2020. https://tinyurl.com/nsvpwt27.
3. Maria Stancu. 2020. "COVID-19 Forces One of the Biggest Surges in Technology Investment in History —KPMG Romania." KPMG. November 23, 2020. https://tinyurl.com/2b7kwpvy.
4. "Report: 91% of Employees Say They're Frustrated with Workplace Tech." VentureBeat. June 10, 2022. https://bit.ly/3Ra1xl2.
5. Eagle Hill Consulting. n.d. "Uncovering ROI: The Hidden Link between Technology Change and Employee Experience." Accessed August 26, 2022. https://bit.ly/3R6mwFC.
6. https://www.salesforce.com/content/dam/web/en_us/www/assets/pdf/platform/salesforce-research-enterprise-technology-trends.pdf.
7. Christian Kelly. n.d. "Power to the People!" Accenture. Accessed August 26, 2022. https://accntu.re/3ASh3wQ.
8. "Harvey Nash. "Everything Changed. Or Did It?" KPMG CIO Survey 2020. December 4, 2020. https://tinyurl.com/yc6xruy5.
9. "New Rackspace Technology Survey of Global IT Leaders Reveals Lack of Confidence, Resources within Their Organizations in Responding to Growing Array of Cyber Threats." 2021. https://bit.ly/3QRRLUN.
10. McKinsey & Company. 2020. "Beyond Hiring: How Companies Are Reskilling to Address Talent Gaps." February 12, 2020. https://mck.co/3TiJkDK.
11. Kevin Townsend. 2019. "With $600 Million Cybersecurity Budget, JPMorgan Chief Endorses AI and Cloud." SecurityWeek.com. https://bit.ly/3con3Up.
12. Peter Bendor-Samuel. 2017. "How to Eliminate Enterprise Shadow IT | Sherpas in Blue Shirts." Everest Group." Everest Group. April 17, 2017. https://bit.ly/3e0rUM7.
13. "PMI Citizen Developer." n.d. Project Management Institute. https://www.pmi.org/citizen-developer.
14. FED. 2017. "How to Harness Citizen Developers to Expand Your AD&D…" Forrester. April 19, 2017. https://bit.ly/3AsAACw.
15. Arnal Dayaratna. 2021. "Quantifying the Worldwide Shortage of Full-Time Developers." IDC: The Premier Global Market Intelligence Company. September 1, 2021. https://bit.ly/3Kqy4B3.
16. Jonathan Frick, KC George, and Julie Coffman. 2021. "The Tech Talent War Is Global, Cross-Industry, and a Matter of Survival." Bain & Company. September 20, 2021. https://bit.ly/3dZZWA6.

17. "Google Career Certificates Fund." n.d. Social Finance. Accessed August 26, 2022. https://bit.ly/3KvQu3G.
18. "Google Introduces 6-Month Career Certificates, Threatening to Disrupt Higher Education with 'the Equivalent of a Four-Year Degree.'" Open Culture. September 5, 2020. https://bit.ly/3Rcu8WT.
19. Abigail Johnson Hess. 2020. "Google Announces 100,000 Scholarships for Online Certificates in Data Analytics, Project Management and UX." CNBC. July 13, 2020. https://tinyurl.com/y7fpppxf.
20. Arnal Dayaratna. 2021. "Quantifying the Worldwide Shortage of Full-Time Developers." IDC. September 1, 2021. https://tinyurl.com/yc8v4j6m.
21. Jen DuBois. 2020. "The Data Scientist Shortage in 2020." QuantHub. April 7, 2020. https://bit.ly/3TmdjLa.
22. Emily Tate. 2017. "Data Analytics Programs Take Off." Inside Higher Ed. March 15, 2017. https://bit.ly/3QRV7XT.
23. Owen Hughes. 2021. "Developers Are Exhausted. Now, Managers Are Worried They Will Quit." TechRepublic. June 29, 2021. https://tek.io/3wxXVlc.
24. Jonathan I. Dingel and Brent Neiman. 2020. "How Many Jobs Can Be Done at Home?" *Journal of Public Economics* 189 (September): 104235. https://bit.ly/3e0h3Bz.
25. Microsoft. 2021. "Microsoft Digital Defense Report." October 2021. https://tinyurl.com/yy5xy52c.
26. "Cyber Security Risks: Best Practices for Working from Home and Remotely." March 30, 2021. https://bit.ly/3pQ6pQS.
27. Computer Economics. 2021. "Application Development Outsourcing on the Rise." Computer Economics—for IT Metrics, Ratios, Benchmarks, and Research Advisories for IT Management. October 11, 2021. https://tinyurl.com/2npjojjd.

>_Chapter 2

1. Wikipedia Contributors. 2018. "Google Sheets." December 21, 2018. https://tinyurl.com/y2d439cq.
2. Javier Soltero. 2020. "Announcing Google Workspace, Everything You Need to Get It Done, in One Location." Google Cloud Blog. October 6, 2020. https://bit.ly/3cjwche.
3. Jordan Novet. 2020. "Google's G Suite Now Has 6 Million Paying Businesses, Up from 5 Million in Feb. 2019." CNBC. April 7, 2020. https://tinyurl.com/r6ynotg.
4. David Thacker. 2019. "5 Million and Counting: How G Suite Is Transforming Work." Google Cloud Blog. February 4, 2019. https://bit.ly/3RgqBXs.
5. *The Today Show.* 2019. "'What Is Internet?' Katie Couric, Bryant Gumbel Are Puzzled." June 20, 2019. https://tinyurl.com/2nctcoes.
6. Funding Universe. n.d. "PayPal Inc. History." https://tinyurl.com/2qydqxeu.

7.   Sissi Cao. 2019. "The Washington Post's Most Valuable Asset Is Now Its Software, Thanks to Jeff Bezos." Observer. October 3, 2019. https://bit.ly/3corjDn.
8.   Cade Metz. 2016. "The Epic Story of Dropbox's Exodus from the Amazon Cloud Empire." *Wired*. March 14, 2016. https://tinyurl.com/gwwg87c.
9.   Ben McCarthy. 2022. "A Brief History of Salesforce." SalesforceBen. August 3, 2022. https://bit.ly/3KqIvod.
10. Julie Bort. 2013. "The Most Controversial and Entertaining Things Larry Ellison Has Ever Said." Business Insider. April 14, 2013. https://bit.ly/3AsIFqP.

>_Chapter 3

1.   TechTarget. 2020. "If You've Got More than One of Them, Are They Computer Mice or Mouses?" WhatIs.com. February 10, 2020. https://tinyurl.com/2g3geq4v.
2.   Dave Farley. 2022. "Will Low Code/No Code Kill Programming Jobs?" YouTube. February 16, 2022. https://tinyurl.com/2jmmzu6p.
3.   Owen Hughes. 2021. "These Old Programming Languages Are Still Critical to Big Companies. But Nobody Wants to Learn Them." TechRepublic. June 30, 2021. https://tinyurl.com/2ff4zrvw.
4.   Google. 2019. "Blockly." Google Developers. 2019. https://developers.google.com/blockly.
5.   FMS. n.d. "Microsoft Access Version Releases, Service Packs, Hotfixes, and Updates History." Accessed August 27, 2022. https://tinyurl.com/2nbpf2r7.
6.   Matthew MacDonald. 2021. "Microsoft Access: The Database Software That Won't Die." Young Coder. October 21, 2021. https://tinyurl.com/y54voymu.
7.   W3. n.d. "Usage Statistics and Market Share of Adobe Dreamweaver." Accessed September 12, 2022. https://tinyurl.com/2jymwnxd.
8.   Squarespace. 2022. "The Essentials of No-Code Website Design." Squarespace. July 21, 2022. https://tinyurl.com/2ejzeva4.
9.   W3. n.d. "Usage Statistics and Market Share of Adobe Dreamweaver." Accessed September 12, 2022. https://tinyurl.com/2jymwnxd.
10. "WordPress Market Share." n.d. Kinsta. Accessed August 26, 2022. https://bit.ly/3ipUSCf.
11. "Bubble Feature Index." n.d. Bubble. Accessed August 26, 2022. https://bubble.io/feature-index.
12. FinSMEs. 2021. "Bubble Raises $100M in Series A Funding." July 27, 2021. https://tinyurl.com/2keorj42.
13. Clive Thompson. 2020. "The New Startup: No Code, No Problem." *Wired*. May 18, 2020. https://tinyurl.com/2h37lce4.
14. Carly Page. 2022. "Microsoft to Block Office Macros by Default Starting July 27." TechCrunch. July 22, 2022. https://tinyurl.com/2y4vlhb9.
15. Andrew Myrick. 2021. "How to Quit All Apps at the Same Time on Your Mac." AppleToolBox. April 14, 2021. https://tinyurl.com/2mv6v9zp.

16. Romain Dillet. 2021. "Apple Is Bringing Shortcuts to the Mac and Starts Transition from Automator." TechCrunch. June 7, 2021. https://tinyurl.com/y6ldvdqe.

17. "Google Trends." n.d. "No-Code Development Platform." Accessed August 26, 2022. https://bit.ly/3CBryWq.

18. Clay Richardson. June 9, 2014. "New Development Platforms Emerge for Customer-Facing Applications." https://bit.ly/3fxP6BV.

19. Rob Marvin. 2014. "How Low-Code Development Seeks to Accelerate Software Delivery." *SD Times*. August 12, 2014. https://bit.ly/3AsCe79.

20. ISO. 2019. "ISO 9407:2019: Footware Sizing—Mondopoint System of Sizing and Marking." June 15, 2019. https://bit.ly/3PRvfdG.

21. Intuit QuickBooks. 2021. "10 Best Apps for QuickBooks Integration." March 15, 2021. https://intuit.me/3pO6Ivi.

22. "Excel Add-In for QuickBooks." n.d. CData Software. Accessed August 26, 2022. https://bit.ly/3QPgQzQ.

23. Intuit QuickBooks. n.d. "QuickBooks App Store." Accessed August 26, 2022. https://intuit.me/3cpodiw.

24. Maddy Osman. 2019. "Wild and Interesting WordPress Statistics and Facts (2022)." Kinsta. June 10, 2019. https://bit.ly/3AtpRri.

25. Anna Fitzgerald. 2022. "20 WordPress Statistics You Should Know in 2022." January 17, 2022. https://bit.ly/3dZnjtE.

26. Elegant Themes. n.d. "The Most Popular WordPress Themes in the World." https://tinyurl.com/kpz7eyh.

27. "Airtable and Slack." n.d. Airtable. Accessed August 26, 2022. https://tinyurl.com/2o6fxtoc.

28. Microsoft. "Real-World Examples of Microsoft Power Automate in Action." Accessed September 30, 2022. https://tinyurl.com/2kompcgn.

29. Phil Simon. 2021. "Automation in Action." airSlate. June 22, 2021. https://tinyurl.com/2je3j9ug.

30. Phil Simon. "r/Airtable—Base Snapshots." Reddit. August 31, 2022. https://tinyurl.com/2geuld2e.

>_Chapter 4

1. Gartner. 2021. "Gartner Forecasts Worldwide Low-Code Development Technologies Market to Grow 23% in 2021." February 16, 2021. https://gtnr.it/3AniXUF.

2. Sarah Frier. 2022. "Zuckerberg's Metaverse to Lose 'Significant' Money in Near Term." Bloomberg. March 25, 2022. https://tinyurl.com/2du2qhwm.

3. Ordinary Things. "Mark Zuckerberg launches Horizon Worlds in France and Spain with an eye-gougingly ugly. . ." Twitter. August 16, 2022. https://tinyurl.com/2l7dkham.

4. Lucas Mearian. 2022. "Low-Code Tools Can Fill a Void Caused by the Great Resignation." Computerworld. April 28, 2022. https://bit.ly/3pO849m.

5. Ron Miller. 2020. "Google Acquires AppSheet to Bring No-Code Development to Google Cloud." TechCrunch. January 14, 2020. https://tcrn.ch/3Tfmg97.

6. Rashi Shrivastava. 2021. "All-In-One Doc Startup Coda Reaches $1.4 Billion Valuation in $100 Million Raise from a Major Pension Fund." *Forbes*. July 8, 2021. https://bit.ly/3CyLd9q.
7. Jared Newman. 2021. "Microsoft Loop Is a Notion Clone for Office Lovers." *Fast Company*. November 2, 2021. https://tinyurl.com/2mcemq77.
8. Glide. n.d. "Glide." https://www.glideapps.com.
9. Draftbit. n.d. "Draftbit: Build the Mobile App You've Always Wanted, Without the Time and Cost." Accessed August 26, 2022. https://draftbit.com/.
10. Natasha Lomas. 2021. "Landbot Closes $8M Series A for Its 'No Code' Chatbot Builder." *TechCrunch*. January 20, 2021. https://tinyurl.com/2g8qtxeh.
11. Swan. n.d. "The Easiest Way to Embed Banking Features into Your Product." Accessed September 5, 2022. https://swan.io.
12. Zac Townsend. 2021. "What the Embedded-Finance and Banking-As-a-Service Trends Mean for Financial Services." McKinsey & Company. March 1, 2021. https://tinyurl.com/embed-ps2.
13. Phil Simon. 2015. "People Need Banking, Not Banks: The Case for Thinking Different." *Wired*. February 25, 2015. https://tinyurl.com/2gusnnj3.
14. Domo. 2020. "Develop Data Apps That Power Any Framework." March 9, 2020. https://tinyurl.com/2hlobodx.
15. Gartner. "Gartner Forecasts Worldwide IT Spending to Exceed $4 Trillion in 2022." October 20, 2021. https://tinyurl.com/y36ccyae.

>_Chapter 5

1. Gartner. "Citizen Developer." Gartner Information Technology Glossary. n.d. Accessed August 26, 2022. https://gtnr.it/3wwRw9y.
2. Dennis Gaughan, Yefim Natis, Gene Alvarez, and Mark O'Neill. n.d. "Future of Applications: Delivering the Composable Enterprise." Gartner. Accessed August 26, 2022. https://gtnr.it/3AiHqdG.
3. Gartner. "Gartner Says the Majority of Technology Products and Services Will Be Built by Professionals Outside of IT by 2024." https://gtnr.it/3APOkZe.
4. Eric Rosenbaum. 2020. "Next Frontier in Microsoft, Google, Amazon Cloud Battle Is over a World Without Code." CNBC. April 1, 2020. https://cnb.cx/3CA6kYR.
5. David F. Carr. 2021. "Gartner: Citizen Developers Will Soon Outnumber Professional Coders 4 to 1." VentureBeat. October 22, 2021. https://bit.ly/3e0xqOM.
6. John Bratincevic. "The Forrester Wave: Low-Code Platforms for Business . . ." Forrester. October 28, 2021. https://bit.ly/3CyKIw4.
7. "The World's Most Important Gathering of CIOs and IT Executives." Gartner IT Symposium. October 19–22, 2020. https://bit.ly/3AOYfOH.
8. U.S. Small Business Administration. 2020. "Frequently Asked Questions." October 2020. https://bit.ly/3csv1vF.
9. Laurie McCabe. 2014. "Why Vendor Definitions of SMB Size Matter." *Laurie McCabe's Blog*. May 5, 2014. https://tinyurl.com/2h4czd4a.

10. CompTIA 2015. "Enabling SMBs with Technology." March 15, 2015. https://tinyurl.com/2hz3grbz.

11. United States Census. 2022. "Business Formation Statistics." August 11, 2022. https://bit.ly/3CBvHts.

12. Clint Boulton. 2021. "CIOs Embrace Business-Led IT amid Tech Democratization." CIO. August 2, 2021. https://bit.ly/3KsjrgW.

13. Cliff Justice and Phil Fersht. 2020. "Enterprise Reboot: Scale Digital Technologies to Grow and Thrive in the New Reality." https://bit.ly/3dZr7Ls.

14. Lucas Mearian. 2022. "Low-Code Tools Can Fill a Void Caused by the Great Resignation." Computerworld. April 28, 2022. https://bit.ly/3pO849m.

15. Conner Forrest. 2015. "'Citizen Developers' Are Ready to Fill the Gaps in Enterprise Applications." TechRepublic. September 29, 2015. https://tek.io/3QV5hY0.

16. Arnal Dayaratna. 2021. "Quantifying the Worldwide Shortage of Full-Time Developers." IDC. September 1, 2021. https://bit.ly/3Kqy4B3.

17. Phil Simon. 2022. "r/nocode: How Would You Describe Your Background?" Reddit. August 16, 2022. https://bit.ly/3CxLlWP.

18. M. Lebens, R. Finnegan, S. Sorsen, and J. Shah (2021). "Rise of the Citizen Developer." *Muma Business Review* 5, no. 12: 101–111. https://doi.org/10.28945/4885.

19. TrackVia. 2014. "The Next Generation Worker: The Citizen Developer." https://bit.ly/3wAN75L.

20. Gene Marks. 2021. "Gen Z Workers Are More Confident, Diverse and Tech-Savvy but Lack Experience | Gene Marks." *Guardian*. December 5, 2021. https://bit.ly/3Kre2GA.

21. Mary Elizabeth Williams. 2021. "Student Debt Is Still Awful. So Why Are We Students Still Taking Out Loans?" Salon. September 6, 2021. https://tinyurl.com/2fo6kdlb.

22. Gallup. 2021. "The American Upskilling Study: Empowering Workers for the Jobs of Tomorrow." September 15, 2021. https://tinyurl.com/2l6jwk57.

23. Beezy. 2022. "2022 Workplace Trends & Insights Report." May 11, 2022. https://tinyurl.com/2kmabx78.

24. Kim Bozzella. 2022. "Are Citizen Developers the New Intrapreneurial Corporate IT Team?" Forbes. August 18, 2022. https://bit.ly/3CyCunA.

>_Chapter 6

1. Stephanie Glen. 2022. "Developer Shortage Fuels Rise in Low-Code/No-Code Platforms." TechTarget. July 14, 2022. https://bit.ly/3KmV3Nq.

2. Mary Branscombe. 2021. "8 Tips for Managing Low-Code Citizen Developers." CIO. December 1, 2021. https://tinyurl.com/2qxd6xwy.

3. Marc Ferranti. 2001. "Gartner Found to Be Lacking in IT-Business Alignment." CIO. November 15, 2001. https://www.cio.com/article/266303/business-alignment-gartner-found-to-be-lacking-in-it-business-alignment.html.

4.   Zapier. 2018. "Email Parser by Zapier." July 2, 2018. https://parser.zapier.com.

5.   Matteo Duò. 2020. "Microsoft Teams vs Slack: Which Collaboration App Is Better?" Kinsta. July 3, 2020. https://tinyurl.com/2zbfonro.

6.   Slack. n.d. "Workflows." Slack API. Accessed August 29, 2022. https://api.slack.com/workflows.

7.   Microsoft. n.d. "Record Form Responses in a Google Sheet." Microsoft Power Automate. Accessed August 26, 2022. https://bit.ly/3Rflib2.

8.   Beezy. n.d. "2021 Digital Workplace Trends & Insights." Accessed September 6, 2022. https://tinyurl.com/2mk38lv5.

9.   ClickUp. n.d. "Zoom Integration with ClickUp." Accessed August 26, 2022. https://clickup.com/integrations/zoom.

10. Notion. n.d. "Timeline View Unlocks High-Output Planning for Your Team." Accessed August 28, 2022. https://tinyurl.com/2qokt85u.

11. Erica Chappell. 2021. "How to Create a Kanban Board in Excel? (with Templates)." *ClickUp* (blog). March 8, 2021. https://tinyurl.com/2n57xxjo.

12. EdrawMax. n.d. "How to Make a Gantt Chart in Google Sheets." https://tinyurl.com/2zb5xq7b.

13. Manasi Sakpal. 2021. "How to Improve Your Data Quality." Gartner. July 14, 2021. https://tinyurl.com/yajckfxa.

14. Marc Ambasna-Jones. 2022. "Why Some Data-Driven Decisions Are Not to Be Trusted." Computer Weekly. March 8, 2022. https://tinyurl.com/y8fwklvn.

15. Airtable. 2022. Season 11, Episode 7, in *BuiltOnAir* (podcast). June 14, 2022. https://tinyurl.com/2euu2vap.

16. Jan-Erik Asplund. 2021. "Zapier: The $7B Netflix of Productivity." March 24, 2021. https://tinyurl.com/2njanpoz.

17. Zapier. n.d. "Post Failed PayPal Charges to Slack." Accessed August 26, 2022. https://bit.ly/3AnnbM1.

18. Phil Simon. 2022. "Episode 69: Next-Level Slack with Christine McHone," in *Conversations About Collaboration* (podcast). August 16, 2022. https://tinyurl.com/2j98qn5k.

19. Clive Thompson. 2020. "The New Startup: No Code, No Problem." *Wired.* May 18, 2020. https://tinyurl.com/2h37lce4.

20. Microsoft. "The Next Great Disruption Is Hybrid Work—Are We Ready?" March 22, 2021. https://tinyurl.com/yg296gkf.

21. Ben Wigert and Jennifer Robison. 2020. "Remote Workers Facing High Burnout: How to Turn It Around." Gallup. October 30, 2020. https://bit.ly/3Tl6sle.

22. McKinsey & Company. 2021. "Employee Burnout Is Ubiquitous, Alarming—and Still Underreported." April 16, 2021. https://mck.co/3e1LTdh.

>_Chapter 7

1.   Joshua Bleiberg and Darrell M. West. 2015. "A Look Back at Technical Issues with Healthcare.gov." Brookings. April 9, 2015. https://tinyurl.com/yxf7xe2u.

2. Mia Hunt. 2022. "Innovation in Government: Lessons from the Netherlands." Global Government Forum. March 5, 2022. https://tinyurl.com/2jxwp64o.
3. Mendix. 2022. "The City of Rotterdam Empowers Development at Scale." August 14, 2022. https://tinyurl.com/2z7j82ny.
4. Eric Katz. 2018. "The Federal Agencies Where the Most Employees Are Eligible to Retire." Government Executive. June 18, 2018. https://tinyurl.com/y3vv3pvt.
5. Microsoft. n.d. "Business Apps | Microsoft Power Apps." https://bit.ly/3TjcdzG.
6. Google Workspace. 2021. "How Kentucky Power Utilized AppSheet." November 5, 2021. https://www.youtube.com/watch?v=TEkezmy5vmQ.

>_**Chapter 8**

1. Fujitsu. 2019. "Fujitsu Highlights Growing Demand for Multi-Cloud Flexibility." February 2019. https://tinyurl.com/2fagjhzf.
2. Lizzy Lawrence. 2022. "Slack or Bust: How Workplace Tools Are Becoming Job Deal-Breakers." Protocol. February 9, 2022. https://tinyurl.com/y7lvonrh.
3. Joel Khalili. 2022. "Microsoft Teams Is About to Pinch One of Slack's Best Features." TechRadar. August 8, 2022. https://tinyurl.com/2r39ot2g.
4. Mary Branscombe. 2021. "8 Tips for Managing Low-Code Citizen Developers." CIO. December 1, 2021. https://tinyurl.com/2qxd6xwy.

>_**Chapter 9**

1. Quote Investigator. n.d. "I Would Spend 55 Minutes Defining the Problem and Then Five Minutes Solving It." Accessed August 30, 2022. https://tinyurl.com/2lajde2v.
2. Sarah Perez. 2021. "Google's AirTable Rival, Tables, Graduates from Beta Test to Become a Google Cloud Product." TechCrunch. June 14, 2021. https://tcrn.ch/3TefK2l.
3. Skillshare. n.d. "Classes Online | Skillshare." www.skillshare.com. https://skl.sh/3CyVtys.
4. H. A. Vlach & C. M. Sandhofer (2012). "Distributing Learning over Time: The Spacing Effect in Children's Acquisition and Generalization of Science Concepts." *Child Development* 83, no. 4: 1137–1144. https://doi.org/10.1111/j.1467-8624.2012.01781.x.
5. Irina V. Kapler, Tina Weston, and Melody Wiseheart. 2015. "Spacing in a Simulated Undergraduate Classroom: Long-Term Benefits for Factual and Higher-Level Learning." *Learning and Instruction* 36 (April): 38–45. https://doi.org/10.1016/j.learninstruc.2014.11.001.
6. Association for Psychological Science. "Back to School: Cramming Doesn't Work in the Long Term." ScienceDaily. https://tinyurl.com/36g3vj. Accessed August 28, 2022.

>_Chapter 10

1. Mike Berg. 2020. "Software Development Life Cycle (SDLC)." Techopedia. November 12, 2020. https://tinyurl.com/y4bx9qgf.
2. IFTTT. n.d. "Connect Your Google Sheets to Twitter with IFTTT." Accessed August 27, 2022. https://tinyurl.com/yjjz5wv6.
3. Simon Black. 2022. "Learn How to Build Your Data Model Rapidly in 3 Easy Steps." Mendix. July 5, 2022. https://tinyurl.com/2h6kaddv.
4. Airtable Community. 2019. "Max Field Limit." October 10, 2019. https://tinyurl.com/2ka9c4vo.
5. Airtable. n.d. "Marketing Campaign Tracking Template—Free to Use." Accessed August 26, 2022. https://bit.ly/3PTF6Q9.
6. Loren Soeiro. 2021. "What's the Curse of Knowledge, and How Can You Break It?" *Psychology Today*. April 28, 2021. https://tinyurl.com/2dlt5o2w.
7. Ernie Smith. 2020. "No Room for Design." Tedium. July 14, 2020. https://bit.ly/3SrN590.
8. "Did MySpace Kill the Potential for Customization on Social Media?" 2020. Tedium. July 14, 2020. https://tinyurl.com/2px82236.
9. Usability.gov. 2019. "User Experience Basics." 2019. https://tinyurl.com/y4bopkhw.
10. Adam Shaari. 2020. "The PowerPoint Slide That Killed 7 Astronauts." Medium. August 16, 2020. https://tinyurl.com/2o35nl3v.
11. Jake Teton-Landis. 2021. "The Data Model behind Notion's Flexibility." Notion. May 18, 2021. https://tinyurl.com/yj9mxwag.

>_Chapter 11

1. Jimmy Jacobson. Personal conversation. September 18, 2022.
2. Harshil Agrawal. 2021. "Building an Expense Tracking App in 10 Minutes." N8n.io. August 11, 2021. https://bit.ly/3cns6Vf.
3. Lucas Mearian. 2022. "Low-Code Tools Can Fill a Void Caused by the Great Resignation." Computerworld. April 28, 2022. https://tinyurl.com/y47xaohx.
4. R. Maheshwari and Kathy Andrews. n.d. "Top 10 Codeless Testing Tools in 2022." Qentelli. Accessed September 9, 2022. https://tinyurl.com/2gptb2fc.
5. Joanna Stern. 2022. "The 3G Shutdown Is Coming—Here's How That Affects You." *Wall Street Journal*. January 14, 2022, sec. Tech. https://+on.wsj.com/3POD4R74.
6. Asana. 2022. "Asana Announces Record First Quarter Fiscal 2023 Revenues." June 2, 2022. https://tinyurl.com/2qs6xct9.
7. Leila Gharani. 2022. "Web Scraping Made EASY with Power Automate Desktop—for FREE & ZERO Coding." YouTube. February 10, 2022. https://bit.ly/3wAJzAh.
8. Yiren Lu. 2020. "Can Shopify Compete with Amazon without Becoming Amazon?" *New York Times*. November 24, 2020. https://tinyurl.com/y2zxkjp4.
9. Google. n.d. "Google UX Design Certificate." https://tinyurl.com/2fmmougd.

10. Martin R. Yeomans, Lucy Chambers, Heston Blumenthal, and Anthony Blake. 2008. "The Role of Expectancy in Sensory and Hedonic Evaluation: The Case of Smoked Salmon Ice-Cream." *Food Quality and Preference* 19, no. 6: 565–73. https://doi.org/10.1016/j.foodqual.2008.02.009.

11. Karim Toubba. 2022. "Notice of Recent Security Incident." *The LastPass* (blog). August 25, 2022. https://tinyurl.com/2jhypdr8.

12. Das Bipro. 2017. "14 Space Discoveries by Amateur Astronomers." RankRed. October 9, 2017. https://tinyurl.com/2ku9wd8u.

13. Mariusz Zydyk. n.d. "Bulletin Board System (BBS)." WhatIs.com. https://tinyurl.com/2oddokvh.

14. Benj Edwards. 2016. "The Lost Civilization of Dial-Up Bulletin Board Systems." *Atlantic*. November 4, 2016. https://tinyurl.com/j7t2la3.

15. w3. 2004. "Tim Berners-Lee, Inventor of the World Wide Web, Knighted by Her Majesty Queen Elizabeth II." July 16, 2004. https://tinyurl.com/2e2y6gzl.

16. Melanie Wachsman. 2021. "Survey: Low-Code and No-Code Platform Usage Increases." ZDNET. July 1, 2021. https://tinyurl.com/2eehhdmd.

>_Chapter 12

1. Eric Rosenbaum. 2020. "Next Frontier in Microsoft, Google, Amazon Cloud Battle Is over a World Without Code." CNBC. April 1, 2020. https://cnb.cx/3CA6kYR.

2. Sarah Gooding. 2018. "Matt Mullenweg Addresses Controversies Surrounding Gutenberg at WordCamp Portland Q&A." WP Tavern. November 10, 2018. https://bit.ly/3PV2Qn1.

3. "Classic Editor." n.d. WordPress.org. Accessed August 26, 2022. https://bit.ly/3pOTgYa.

4. Andrew Ofstad. 2021. "Putting the Power of Software Development in the Hands of Everyone." Protocol. November 10, 2021. https://bit.ly/3PRXRDE.

5. Accenture. 2022. "Q4 FY22 (Ended August 31, 2022)." https://tinyurl.com/2gw2arng.

6. Phil Simon. 2020. "Using Slack to Ghostwrite a Book: A Case Study." February 24, 2020. https://tinyurl.com/2mqflpt4.

Epilogue

1. Manish Singh. 2021. "Notion Acquires India's Automate.io in Push to Accelerate Product Expansion." TechCrunch. September 8, 2021. https://tinyurl.com/yhcg39z7.

2. Okta. 2021. "Businesses @ Work." March 23, 2021. https://tinyurl.com/yhbn8efq.

3. Okta. n.d. "Okta Workflows." Accessed August 26, 2022. https://bit.ly/3Torr6K.

4. Pendo. n.d. "Software That Makes Your Software Better." https://www.pendo.io.

5. Rina Diane Caballar. 2021. "Programming by Voice May Be the Next Frontier in Software Development." IEEE Spectrum. March 22, 2021. https://bit. ly/3e0zjuQ.

6. Ron Miller. "Serenade Snags $2.1M Seed Round to Turn Speech into Code." TechCrunch. November 23, 2020. https://tcrn.ch/3Rise7n.

7. Jason Wong. 2019. "The Importance of Citizen Development and Citizen IT." October 10, 2019. https://gtnr.it/3TocvWe.